Electromechanical Machinery Theory and Performance

Electromechanical Machinery Theory and Performance

Thomas Ortmeyer
Clarkson University, Potsdam, New York, USA

IOP Publishing, Bristol, UK

Supplementary material for this book is available from http://iopscience.iop.org/book/978-0-7503-1662-0.

ISBN 978-0-7503-1662-0 (ebook)
ISBN 978-0-7503-1660-6 (print)
ISBN 978-0-7503-1661-3 (mobi)

DOI 10.1088/978-0-7503-1662-0

Version: 20180801

IOP Expanding Physics
ISSN 2053-2563 (online)
ISSN 2054-7315 (print)

British Library Cataloguing-in-Publication Data: A catalogue record for this book is available from the British Library.

Published by IOP Publishing, wholly owned by The Institute of Physics, London

IOP Publishing, Temple Circus, Temple Way, Bristol, BS1 6HG, UK

US Office: IOP Publishing, Inc., 190 North Independence Mall West, Suite 601, Philadelphia, PA 19106, USA

Contents

Preface

This book is intended as an introduction to electromagnetic machines. In particular, transformers, synchronous machines and induction machines are covered. The book also provides a brief introduction to single phase and three phase pulse width modulation inverters, and to variable alternating current (AC) motor drives. The book begins with a brief introduction to energy sources and uses, followed by a review of single phase and three phase AC power and magnetic circuits. The book is intended for those who have a background in physics and mathematics typical of first and second year engineering students, plus an engineering course in electric circuits.

The book was developed from course notes for the Energy Conversion course taught at Clarkson University. This is a one semester course, and covers most of the material in the book apart from three phase transformers. The book is also written to be accessible for self-study to practicing engineers from all disciplines who have an interest in this topic.

The focus of the course is on the nature of these machines, and their application. It therefore includes some material not common in many textbooks, while omitting other material that is less relevant today. Each chapter ends with both questions and problems. The majority of the questions involve looking up material that is commonly found on the internet. The problems are numerical in nature, and follow the material and examples in the chapters. Exercises are included that encourage the use of computers to perform repetitive calculations. Several Excel files are included that can be used for some of these, and all of these can be readily solved using easily available numerical software products. These can be downloaded from http://iopscience.iop.org/book/978-0-7503-1662-0. Instructors can contact the author at tortmeye@clarkson.edu for information on instructional support materials.

I thank my wife Ann for her patience during the many long evenings working on this manuscript. I also thank the many students who took this course from me over the past several years. Finally, I acknowledge Jefferson Electric, and thank them for providing and allowing us to include the transformer nameplate in chapter 3.

Tom Ortmeyer
Clarkson University
April, 2018

Author biography

Thomas H. Ortmeyer

Tom Ortmeyer began his career at Commonwealth Edison Company in the Operations Analysis Department. He then returned to Iowa State University to pursue graduate studies. Upon completion of his PhD degree, he joined the Electrical and Computer Engineering department faculty at Clarkson University, where he is currently Research Professor. At Clarkson, he has taught many courses in the power engineering area. He is currently involved in research in the areas of power distribution, distributed generation interconnection, microgrids, and power system protection. He is a Life Fellow of IEEE.

IOP Publishing

Electromechanical Machinery Theory and Performance

Thomas Ortmeyer

Chapter 1

Energy and power

Efficient conversion of energy from one form to another is an important component of modern life. This book discusses electromagnetic conversion of transformers and rotating machines.

Transformers convert alternating currents (AC) from one voltage level to another. They can also provide electrical isolation between circuits. Electromechanical machines convert energy from electrical to mechanical or mechanical to electrical. Most, but not all, of these machines are rotating. Over 90% of our electric power is generated by rotating machines. The voltage of the generated power is stepped up to levels where it is efficient to transmit the energy over long distances. Transformers then step the voltage down so it can be distributed and used in a wide variety of applications.

These machines depend on the ferromagnetic properties of iron and other ferromagnetic materials. The high permeability of iron allows the high energy density transformations needed for these machines.

This book presents a basic introduction to ferromagnetic circuits. It then discusses the basics of transformers. The book then shifts focus to rotating machines, and covers the basic performance characteristics of synchronous and induction machines that are directly connected to the AC power system. This is followed by an introduction to three phase pulse width modulated (PWM) converters. The book concludes with material on variable speed operation of AC machines.

1.1 Energy

1.1.1 Energy sources

The ready availability of energy is a key feature of modern life. Energy is available from a variety of sources, that include

- Fossil
    - Coal
    - Petroleum
    - Natural gas

doi:10.1088/978-0-7503-1662-0ch1

- Non-fossil
  - Hydropower
  - Wind
  - Solar
  - Geothermal
  - Biomass
  - Nuclear fission

In 2013, total energy consumption in the United States was 97.1 quadrillion BTU [1]. The sources of this energy are shown in table 1.1.

The energy reported in table 1.1 is given in terms of quadrillion BTU's (quads) per year. In general, units of energy are less familiar to us than other units, including current, voltage and power. Another important note in table 1.1 is that these units are given in terms of a time period—in this case quads per year.

Another common measure of national and world energy units is exajoules (10^{18} joules) per year, a common measure of energy. The relationship between these two units of energy is

$$1 \text{ quad} = 1.055 \text{ EJ}$$

Electrical energy is more commonly measured in units of watt-hours, kilowatt-hours or megawatt-hours. For example, 1 kilowatt-hour of electric energy purchased from the power grid costs in the neighborhood of 10 cents, depending on where it is purchased.

1 joule is a watt-second, where watt is the MKS unit of power and joule is the MKS unit of energy. The conversion from joules to watt-hours is then the conversion from watt-seconds to watt-hours,

$$1 \text{ Wh} = 3600 \text{ J} \tag{1.1}$$

Table 1.1. Energy sources, United States, 2015.[1]

Energy source	Amount (quadrillion BTU per year)	Amount by percentage
Liquid fuels and other petroleum	36.5	37.7
Natural gas	28.2	29.2
Coal	15.5	16.0
Nuclear/uranium	8.3	8.6
Hydropower	2.3	2.4
Biomass	2.8	2.9
Solar (photovoltaic and solar thermal)	0.2	0.2
Wind	1.8	1.9
Other	1.1	1.1
Total	96.7	100%

[1] 2016 Annual energy outlook, US Energy Information Administration.

As can be seen, a kilowatt-hour is a more convenient unit of energy for electric power bills than is a joule. While this book generally follows MKS units, the units of energy are generally expressed in kilowatt-hours or megawatt-hours.

Energy is the integral of power over time,

$$W = \int_{t_1}^{t_2} P(t)dt \tag{1.2}$$

If you perform this integral with P in watts and t in seconds, the resulting energy is in joules. If you perform the integral with P in kilowatts and time in hours, the result of the integral is in units of kilowatt-hours.

1.1.2 Electrical energy

Electricity is not a source of energy. It is a convenient and competitive method for transporting energy. It is also the most flexible form of energy for energy users. Table 1.2 shows the main energy transportation technologies in current use. Each of these technologies has both advantages and disadvantages. Each can be controversial—with some controversies reaching national and even international levels.

Table 1.3 shows the forms of energy used at the point of consumption, by sector. Table 1.4 shows several prominent applications that use energy, and the primary form of energy that is often used to power these applications.

Table 1.2. Technologies in common usage for transporting energy.

Energy transportation technology	Primary energy sources used
Electricity	Natural gas, coal, nuclear, hydro, biomass, solar, wind, landfill gas, other
Pipelines, natural gas	Natural gas, coal gasification
Pipelines, liquid fuel	Petroleum, biofuels
Rail	Coal, petroleum, natural gas
Road	Petroleum
Ship, barge	Coal, petroleum, natural gas

Table 1.3. Delivered energy consumed at point of use, by sector, quads/year (2015).

	Residential	Commercial	Industrial	Transportation
Electricity	4.78	4.64	3.27	0.03
Natural gas	4.77	3.32	9.38	0.90
Liquid fuel and other petroleum (transportation total includes ethanol based fuels)	0.93	0.66	8.07	27.14
Coal	0.0	0.0	1.34	0.0
Renewable	0.44	0.14	1.48	0.0
Sector total	9.44	8.81	24.33	28.13

Table 1.4. Energy end use applications and primary form of energy consumed.

Application	Primary energy form
Lighting	Electricity
Space heating	Natural gas
Space cooling	Electricity
Water heating	Natural gas
Stationary rotational power (electric or gasoline motors for elevators, mills, etc)	Electricity
Transportation	Petroleum
Electronics, communications and computing	Electricity

Table 1.5. Energy sources for electric power generation.

Primary energy source	Billion kilowatt-hours per year, 2011	Billion kilowatt-hours per year, 2012	Billion kilowatt-hours per year, 2013	Billion kilowatt-hours per year, 2015
Coal	1688	1478	1550	1355
Petroleum	26	18	22	26
Natural gas	804	1000	894	1348
Renewable sources	476	458	483	564
Nuclear	790	769	789	798

Table 1.5 shows the primary energy source for electric power in the United States in 2011–2015. In recent years, the proportion of electricity generated from coal has been declining, while the proportion generated from natural gas and renewables has been increasing. Approximately half of the renewable total in 2011 was from traditional hydropower. This level of generation has been stable in the recent past, and likely will remain at about this level in the foreseeable future. The remainder of renewable generation comes from solar, wind, geothermal and biomass sources. These renewable technologies have recently started to become significant, and several of these will continue to grow in the coming years.

1.2 Power—single phase

Power is the rate at which energy is generated, transported, or consumed,

$$P = \frac{dW}{dT} \tag{1.3}$$

Equation (1.3) is a restatement of equation (1.2), but nevertheless is worth repeating. The units of power are most commonly watts. Any power calculations done in this book will be in terms of watts (or kilowatts or megawatts). Very occasionally, we will convert results to a different unit. An example of this is horsepower, where

$$1 \text{ horsepower} = 746 \text{ W} \tag{1.4}$$

The power rating of many motors sold in the United States historically has been stated in terms of horsepower. Increasingly, however, we are seeing motors rated in watts in the US. Motor ratings in watts is nearly universal in the rest of the world.

Example 1.1. An incandescent light bulb draws 100 W when it is turned on. Find the energy consumed in a day, when it is on for 10 h that day.

From equation (1.3), the energy drawn is

$$W_{\text{bulb}} = \int_{t=0}^{t_{\text{final}}} P_{\text{bulb}} dt$$

When the units of time are seconds,

$$W_{\text{bulb}} = \int_{t=0}^{10 \cdot 60 \cdot 60} 100dt = 3.6 \times 10^6 \text{J} = 3.6 \text{ MJ}$$

When the units of time are hours,

$$W_{\text{bulb}} = \int_{t=0}^{10} 100dt = 1000 \text{ Wh} = 1.0 \text{ kWh}$$

Example 1.2. A 10 horsepower electric motor supplies rated load to a mechanical load for 15 min out of every hour for an 8 h period.

(a) What is the mechanical energy supplied to the load?

$$P_{\text{motor}} = 10 \text{ horsepower} \cdot \frac{746 \text{ W}}{1 \text{ horsepower}} = 7460 \text{ W}$$

As the power draw is constant, equation (1.2) simplifies to

$$W_{\text{motor}} = P_{\text{motor}} \cdot t_{\text{ON}} = 7460 \text{ W} \cdot 8 \text{ h} \cdot 0.25 \frac{\text{hours}}{\text{hour}} = 14920 \text{ Wh}$$

This can be stated as 14.92 kWh. Note that the motor being on for a quarter hour each hour for 8 h is a total on time of 2 h, and that 7460 W = 7.46 kW. This calculation can be made as

$$W_{\text{motor}} = 7.46 \text{ kW} \cdot 2 \text{ h} = 14.92 \text{ kWh}$$

(b) If a kWh of electricity costs 15 cents, what will be the cost of running the motor for a day?

$$\text{Cost} = \$0.15 \cdot 14.92 \text{ kWh} = \$2.24$$

(c) What is the cost of a joule of electricity when it costs 15 cents per kWh?

1 kWh = 3.6×10^6 J, so the cost of 1 J of electricity would be

$$\text{cost per joule} = 15\,\frac{\text{cents}}{\text{kWh}}\,\frac{1\ \text{kWh}}{3.6\times 10^{6}\ \text{J}} = 4.17\times 10^{-6}\ \text{cents}$$

This shows that the kWh is a more practical unit for electric power than is the joule.

1.2.1 Power in sinusoidal single phase electrical circuits

Figure 1.1 shows a simple electrical circuit that is a good representation of many single phase power circuits. In this circuit, power flows from the sinusoidal voltage source $\bar{V}_s$ (on the left) through impedances $\bar{Z}_S$ and $\bar{Z}_L$ to a load, which in this case is represented by a 'source' $\bar{V}_L$. In practice, loads can look like voltage sources, impedances, current sources, or a combination of these. The current returns through a neutral path that is represented as having zero impedance. In most cases, this neutral path is grounded at one or more points.

In this case, we are interesting in sinusoidal AC operation, and the voltages and currents are represented in the phasor domain.

The phasors are defined in terms of their root mean square (RMS) values. RMS values are also referred to as the effective value of a quantity, as an RMS voltage delivers that same power to a resistor as does a dc voltage. The magnitudes of AC voltages or currents are most commonly stated in terms of their RMS values. For example, we think of the nominal output voltage of a wall plug being 120 V. This is the RMS value of that voltage. Similarly, if we are referring to a 20 A circuit feeding the wall plug, this is 20 A RMS. However, occasionally the magnitudes of AC quantities are stated as peak values. This can cause confusion, so if there is any doubt it is useful to ask for clarification. It is also worth stating that one reason for using RMS values for the magnitude of voltages or currents is that the RMS value is more useful than peak values in reporting the magnitude when there is distortion in the sinusoidal waveforms.

In the time domain, the voltages in figure 1.1 can be written as

$$\begin{aligned} v_s(t) &= \sqrt{2}\,V_s\cos(\omega_e t + \delta_s) \\ v_x(t) &= \sqrt{2}\,V_x\cos(\omega_e t + \delta_x) \\ v_L(t) &= \sqrt{2}\,V_L\cos(\omega_e t + \delta_L) \end{aligned} \tag{1.5}$$

Note that the frequency ω_e is the same throughout the circuit.

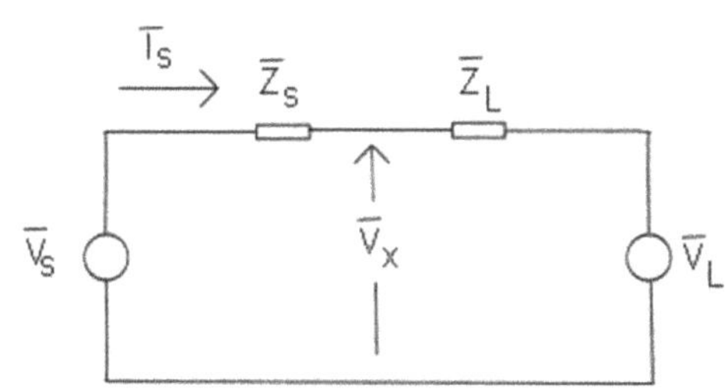

Figure 1.1. Simple AC circuit with phasor notation.

In the phasor domain, these voltages are written as:

$$\begin{aligned} \bar{V}_s &= V_s \underline{/\delta_s} \\ \bar{V}_X &= V_X \underline{/\delta_X} \\ \bar{V}_L &= V_L \underline{/\delta_L} \end{aligned} \tag{1.6}$$

The notation used in this book is that phasor values such as the voltage $\bar{V}_X$ (with overbar) is the complex phasor quantity, and the voltage V_X (without overbar) is the magnitude of the phasor quanity, a real number.

Similarly, the source current in figure 1.1 can be written in either the time domain or the phasor domain as

$$\begin{aligned} i_s(t) &= \sqrt{2}\, I_s \cos(\omega_e t + \phi_s) \\ \bar{I}_s &= I_s \underline{/\phi_s} \end{aligned} \tag{1.7}$$

In this circuit, source and load current are the same.

The instantaneous power flow at any point in this circuit is the product of the time domain circuit voltage at that point and the time domain current flowing at that point. At the source, for example, the instantaneous power is

$$p_s(t) = [\sqrt{2}\, V_s \cos(\omega_e t + \delta_s)][\sqrt{2}\, I_s \cos(\omega_e t + \phi_s)] \tag{1.8}$$

An equivalent expression of this power is

$$p_s(t) = [V_s I_s][\cos(\delta_s - \phi_s) + \cos(2\omega_e t + \delta_s + \phi_s)] \tag{1.9}$$

From equation (1.9), it is clear that the average power at this location in the circuit is

$$P_s = [V_s I_s][\cos(\delta_s - \phi_s)] \tag{1.10}$$

In this equation, the upper case P signifies that it is the average power, while the lower case p in equation (1.9) is used to represent the instantaneous power. The average power flow at any of the points on this circuit can be found by substituting the appropriate voltage and current into equation (1.10).

In steady state phasor analysis of AC circuits, the passive elements are represented as impedances. The impedances of ideal resistors, inductors, and capacitors are

$$\begin{aligned} \bar{Z}_R &= R \\ \bar{Z}_L &= j\omega_e L \\ \bar{Z}_C &= \frac{1}{j\omega_e C} = \frac{-j}{\omega_e C} \end{aligned} \tag{1.11}$$

From this equation, it can be seen that the impedance of a resistor has a real value, the impedance of an inductor has a positive imaginary value, and the impedance of a capacitor has a negative imaginary value. Also, the impedances of the inductor and capacitor are a function of the frequency applied to the circuit, while the impedance of the ideal resistor is not. Complex impedances result from series and parallel connections of these elements.

In most North American power systems, the frequency of the circuit is fixed at a (nearly) constant 60 Hz. Engineers who work constantly with constant frequency power systems find it convenient to define the term 'reactance' to represent the imaginary part of an impedance. The symbol 'X' is used to represent reactance. Remember that reactance X itself is a real number—the impedance

$$\overline{Z} = R + jX = 4 + j9 \text{ ohms} = 9.85\underline{/66^\circ} \text{ ohms} \tag{1.12}$$

has a resistance of 4 ohms and a reactance of 9 ohms. It also is said to have a magnitude of 9.85 ohms, and angle of 66°.

When expressed in the phasor domain, the steady state solution of figure 1.1 is straightforward,

$$\overline{I}_s = \frac{\overline{V}_s - \overline{V}_L}{\overline{Z}_s + \overline{Z}_L} \tag{1.13}$$

The voltage $\overline{V}_x$ can be found readily, and the power flow at source, mid-point, and load follows. It is emphasized that the power equation (1.10) uses the magnitudes and not the phasor quantities in computing the power.

1.2.2 Reactive power and apparent power

The power flow in an AC circuit is expressed in watts, and represents the rate of flow of energy in that circuit. In equation (1.10), the average power flow at a point in a circuit is given in terms of the voltage and current magnitudes and phase angles.

It has been standard and longstanding industry practice to define two additional quantities for AC circuits—the apparent power and the reactive power.

The apparent power flowing at some point in a single phase circuit is defined as the product of the magnitude of the voltage and the magnitude of the current. It is designated with the symbol 'S', and has units of volt-amps (VA). The apparent power flowing at point x in figure 1.1 is therefore

$$S_x = V_x I_s \tag{1.14}$$

The reactive power flowing at the same point in this circuit is defined as

$$Q_x = [V_x I_s][\sin(\delta_x - \phi_x)] = S_x \sin(\delta_x - \phi_x) \tag{1.15}$$

Reactive power is designated by the symbol 'Q' and has units of volt-amps reactive, or VARs.

While both apparent power and reactive power appear to have the same units as the real power (which is watts), they do not represent the average flow of real power. Therefore, it is not appropriate to state their units as watts, even though real power, apparent power, and reactive power all are calculated as the product of voltage and current.

Apparent power and reactive power do have practical value, however. The capacity of generators, lines and transformers is based on the apparent power flow, not the real power flow. The capability of a power system to transmit real power is therefore limited by the amount of reactive power flowing in the circuit. The power factor has been defined as the ratio of real power flow to apparent power flow,

$$\text{power factor } pf_x = \frac{P_x}{S_x} = \cos(\delta_x - \phi_x) = \cos(\theta_x) \tag{1.16}$$

The value of the power factor is often expressed as a percentage rather than a straight ratio. If a given line has power flow with a power factor of 50%, that line can only transmit half the watts that could be delivered if the power factor were to be 100%.

Note that in equation (1.16), the angle θ is used to represent the angle by which current lags the voltage.

Equations (1.10), (1.14) and (1.15) have the same relationship as does a right triangle. It is therefore possible (and often helpful) to diagram their relationship as shown in figure 1.2. This is termed the power triangle. It is immediately clear from the power triangle that

$$\begin{aligned} S_x &= \sqrt{P_x^2 + Q_x^2} \\ \theta_x &= \cos^{-1}\frac{P_x}{S_x} = \tan^{-1}\frac{Q_x}{P_x} \end{aligned} \tag{1.17}$$

Many circuits books include additional information on the theory and physical meaning of apparent power and reactive power. For the material covered in this book, the understanding given by the power triangle and the associated equations is sufficient for our purposes.

Example 1.3. A single phase power circuit is shown in figure 1.3. In this case, there is a single load that has a constant voltage magnitude of 112 V. Power flows from the source to an intermediate point X, and then to the load L. Point X for example could

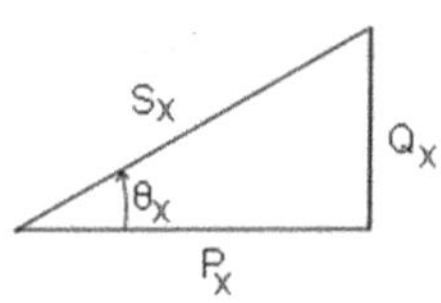

Figure 1.2. Power triangle for point *X*.

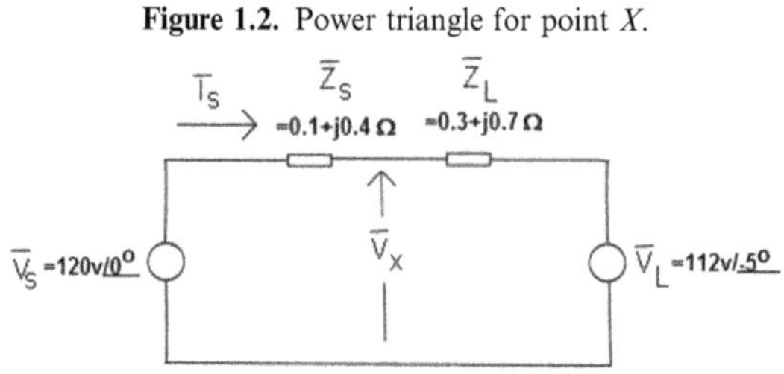

Figure 1.3. Single phase power circuit with load voltage known.

be the point where the power grid connects to the consumer—which is the point where the energy meter will be connected.

The current in the circuit is

$$\overline{I}_s = \frac{\overline{V}_s - \overline{V}_L}{\overline{Z}_s + \overline{Z}_L} = \frac{120\ \text{V}\underline{/0^\circ} - 112\ \text{V}\underline{/-5^\circ}}{(0.1 + j0.4)\Omega + (0.3 + j0.7)\Omega}$$
$$= (10.30 - j3.92)\ \text{A} = 11.2\ \text{A}\angle\underline{/-20.8^\circ}$$

For this circuit, this same current $\overline{I}_s$ will flow from the source, through point X, and into the load. At an angle of −20.8°, this current lags both the source and the load voltages.

The voltage $\overline{V}_X$ can then be found to be

$$\overline{V}_X = \overline{V}_s - \overline{Z}_s \cdot \overline{I}_s = 120v\underline{/-0^\circ} - (0.1 + j0.4)\Omega \cdot 11.2\,A\underline{/-20.8^\circ} = 117.5\ \text{V}\underline{/-1.8^\circ}$$

Real power: The direction of real power flow in the circuit is from source to load. The real power flows at points S, X, and L are

$$P_S = V_S \cdot I_S\cos(\delta_S - \phi_S) = 120\ \text{V} \cdot 11.2\ \text{A} \cdot \cos(0^\circ - (-20.8^\circ)) = 1236\ \text{W}$$
$$P_X = V_X \cdot I_S\cos(\delta_X - \phi_S) = 117.5\ \text{V} \cdot 11.2\ \text{A} \cdot \cos(-1.8^\circ - (-20.8^\circ)) = 1224\ \text{W}$$
$$P_L = V_L \cdot I_S\cos(\delta_L - \phi_S) = 112\ \text{V} \cdot 11.2\ \text{A} \cdot \cos(-5^\circ - (-20.8^\circ)) = 1187\ \text{W}$$

Note that the voltage and current values in these equations are magnitude rather than phasor quantities. The notation we are using for voltage angle at node k is δ_k, and for current in component k is ϕ_k.

Apparent power: The apparent power flows at the three points in this circuit are:

$$S_S = V_S \cdot I_S = 120\ \text{V} \cdot 11.2\ \text{A} = 1322\ \text{VA}$$
$$S_X = V_X \cdot I_S = 117.5\ \text{V} \cdot 11.2\ \text{A} = 1295\ \text{VA}$$
$$S_L = V_L \cdot I_S = 112\ \text{V} \cdot 11.2\ \text{A} = 1234\ \text{VA}$$

Reactive power: The reactive power flows at the three points in this circuit are:

$$Q_S = V_S \cdot I_S\sin(\delta_S - \phi_S) = 120\ \text{V} \cdot 11.2\ \text{A} \cdot \sin(0^\circ - (-20.8^\circ)) = 470\ \text{VARs}$$
$$Q_x = V_X \cdot I_S\sin(\delta_X - \phi_S) = 117.5\ \text{V} \cdot 11.2\ \text{A} \cdot \sin(-1.8^\circ - (-20.8^\circ)) = 422\ \text{VARs}$$
$$Q_L = V_L \cdot I_S\sin(\delta_L - \phi_S) = 112\ \text{V} \cdot 11.2\ \text{A} \cdot \sin(-5^\circ - (-20.8^\circ)) = 336\ \text{VARs}$$

Note that the apparent power is always positive, and does not indicate the direction of flow. The real and reactive powers, however, can be either positive or negative and do indicate the direction of flow. The positive direction for this flow is set by the direction chosen for the line current flow. In this example, real and reactive power flow into the load. The line resistances cause losses, so the real power flow at points X and S are higher due to these losses. Similarly, the line reactances consume reactive power, and VAR flow at points X and S are increased in order to support these line 'losses' plus the reactive power drawn by the load.

Source power triangle: The power triangle for the source is shown in figure 1.4.

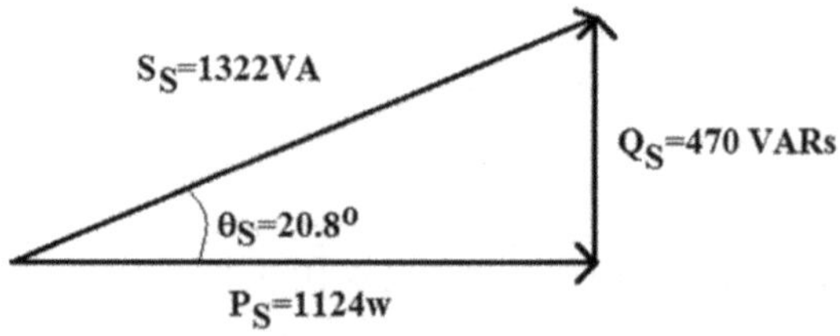

Figure 1.4. Power triangle for the source in example 1.3.

1.3 Three phase power

Three phase power is used in situations where moderate to large amounts of electric power are involved (10–100 of kilowatts and above). Three phase sources and loads are connected in either wye or delta. The wye connection is the more basic of these connections, and we will start by analyzing a system with a wye connected source and load.

1.3.1 A balanced wye connected power system

Figure 1.5 shows a simple power system with a wye connected source and a wye connected load. In this diagram, the notation E (electromotive force) is used for the voltage source. On the system, the voltages are denoted with the notation V, with subscripts showing the location of the voltage. Therefore, the voltage between the point A and the point N is $\bar{V}_{AN}$. The voltage drop on the line in the A phase is $\bar{V}_{AA'} = \bar{Z}_{Line}\bar{I}_A$. When a voltage is stated with respect to ground, the second subscript is often omitted. So the voltage from point A to ground is often stated as $\bar{V}_A$. In this case, $\bar{V}_A = \bar{V}_{AN}$ as the neutral point N is connected directly to ground.

The balanced source has three voltages,

$$\begin{aligned} \bar{E}_A &= E_s \underline{/\delta_s} \\ \bar{E}_B &= E_s \underline{/\delta_s - 120^\circ} \\ \bar{E}_C &= E_s \underline{/\delta_s - 240^\circ} \end{aligned} \tag{1.18}$$

These three source voltages have equal magnitudes and are separated in time by 120°—B phase lags A phase by 120°, and C phase lags B by 120°. The A phase then lags C by the same 120°, creating symmetry among the three phases. This symmetry has several useful properties, which will be shown at various points in this text.

Line currents: The system shown in figure 1.5 has equal source impedances $\bar{Z}_s$, line impedances $\bar{Z}_{Line}$, and load impedances $\bar{Z}_{wye}$. The system has a zero impedance neutral path, connecting the source neutral point N and the load neutral point N′. Because this path has zero impedance, each phase of the circuit can be treated independently. The A phase current is

$$\bar{I}_A = \frac{\bar{E}_A}{\bar{Z}_s + \bar{Z}_{Line} + \bar{Z}_{wye}} = I_A \underline{/\phi_s} \tag{1.19}$$

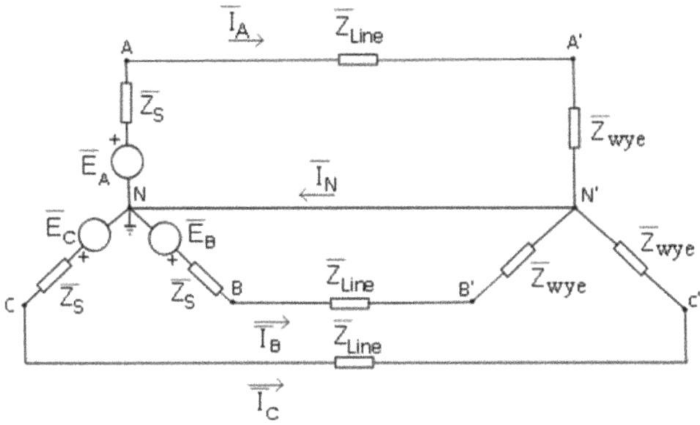

Figure 1.5. Balanced three phase circuit with wye connected source and wye connected load.

Due to the symmetry of the voltages and the equal source, line and load impedances, B and C phase currents will be

$$\overline{I}_B = \frac{\overline{E}_B}{\overline{Z}_s + \overline{Z}_{Line} + \overline{Z}_{wye}} = \frac{(\overline{E}_A)(1\underline{/-120°})}{\overline{Z}_s + \overline{Z}_{Line} + \overline{Z}_{wye}} = I_A\underline{/\phi_s - 120°}$$
$$\overline{I}_C = \frac{\overline{E}_C}{\overline{Z}_s + \overline{Z}_{Line} + \overline{Z}_{wye}} = \frac{(\overline{E}_A)(1\underline{/-240°})}{\overline{Z}_s + \overline{Z}_{Line} + \overline{Z}_{wye}} = I_A\underline{/\phi_s - 240°} \tag{1.20}$$

These currents are also balanced—equal magnitudes and separated in time by 120°. From figure 1.5, the phase currents flow from source to load, meet at the load neutral point N′, and then travel back through the neutral to the source. From Kirchhoff's current law,

$$\overline{I}_N = \overline{I}_A + \overline{I}_B + \overline{I}_C \tag{1.21}$$

Due to the symmetry of this balanced three phase circuit,

$$\overline{I}_N = \overline{I}_A(1 + 1\underline{/-120°} + 1\underline{/-240°}) = 0 \tag{1.22}$$

The neutral current in this three phase balanced circuit is therefore zero. This is one of the important benefits of three phase balanced circuits, and it holds for all balanced three phase circuits, regardless of the complexity of the circuit or connection of source or load. It is also worth noting that the voltage drop from N′ to N will be zero in the balanced case, even when there might be some impedance between these two points.

Voltages—line to neutral and line to line: Of course, circuit voltages are also of interest. At any given location, it is common to report voltages as either line to neutral or line to line. In the balanced wye circuit of figure 1.5, these terms are unambiguous. The A phase to neutral voltage at the load is $\overline{V}_{A'N}$, and the line to neutral voltage at the source terminals is $\overline{V}_{AN}$ Note that there is no voltage drop on the neutral from N′ to N, as the circuit model has no impedance in the neutral. Also, note that we will use double subscript notation on the voltage to indicate that this is a voltage from point A′ to point N. Even on a simple diagram like shown in figure 1.5, the figure would become cluttered and confusing if each possible voltage was to be specifically labeled on the figure.

Through balanced three phase circuit symmetry, the line to neutral voltages at the source terminals are

$$\begin{aligned} \bar{V}_{AN} &= V_t \underline{/\delta_t} \\ \bar{V}_{BN} &= V_t \underline{/\delta_t - 120°} \\ \bar{V}_{CN} &= V_t \underline{/\delta_t - 240°} \end{aligned} \tag{1.23}$$

In practice, by longstanding convention, when referring to a voltage from a point to ground, the reference to ground in the subscript can be omitted. In this case, the neutral point N is connected directly to ground, so they are at the same potential. Therefore, the following three quantities all refer to the same voltage in this case, as the ground and neutral points are common in this figure,

$$\bar{V}_A = \bar{V}_{AG} = \bar{V}_{AN} \tag{1.24}$$

In a three phase wye system, it is often desirable to know the line to line voltage at some location in the system. The voltage between phases A and B in figure 1.5 is

$$\bar{V}_{AB} = \bar{V}_{AN} - \bar{V}_{BN} = \sqrt{3}\, V_T \underline{/\delta_T + 30°} \tag{1.25}$$

Similarly, the B to C and C to A voltages are

$$\begin{aligned} \bar{V}_{BC} &= \bar{V}_{BN} - \bar{V}_{CN} = \sqrt{3}\, V_T \underline{/\delta_T - 90°} \\ \bar{V}_{CA} &= \bar{V}_{CN} - \bar{V}_{AN} = \sqrt{3}\, V_T \underline{/\delta_T + 150°} \end{aligned} \tag{1.26}$$

As is the case with line to neutral voltages, the three line to line voltages in balanced systems have equal magnitude and are separated by 120°. Further note that the magnitude of the line to line voltage is $\sqrt{3}$ greater than the magnitude of the line to neutral voltage, and that the angle of the line to line voltage is different than the angle of the line to neutral voltage by 30°. These relationships are shown graphically in figure 1.6, for the case where $\delta_T = 0$.

In figure 1.6(a), the line to line voltages are shown as the vector difference between two of the line to neutral voltages. In figure 1.6(b), the line to line voltage phasors are drawn with a common reference point, with the magnitude and angle retained from (a). Note that in (b), the common point has no physical meaning, while in (a), the common point of the line to neutral phasors does represent the physical neutral point of a wye connection.

Example 1.4. In the balanced three phase wye–wye circuit shown in figure 1.5, the A phase source voltage is $\bar{E}_A = 7200\ \text{V}\ \underline{/0°}$. The source impedance is $\bar{Z}_s = 0.25 + j1.0\ \Omega$, the line impedance is $\bar{Z}_{Line} = 0.6 + j0.9\ \Omega$, and the wye connected load impedance is $\bar{Z}_{wye} = 35 + j20\ \Omega$.

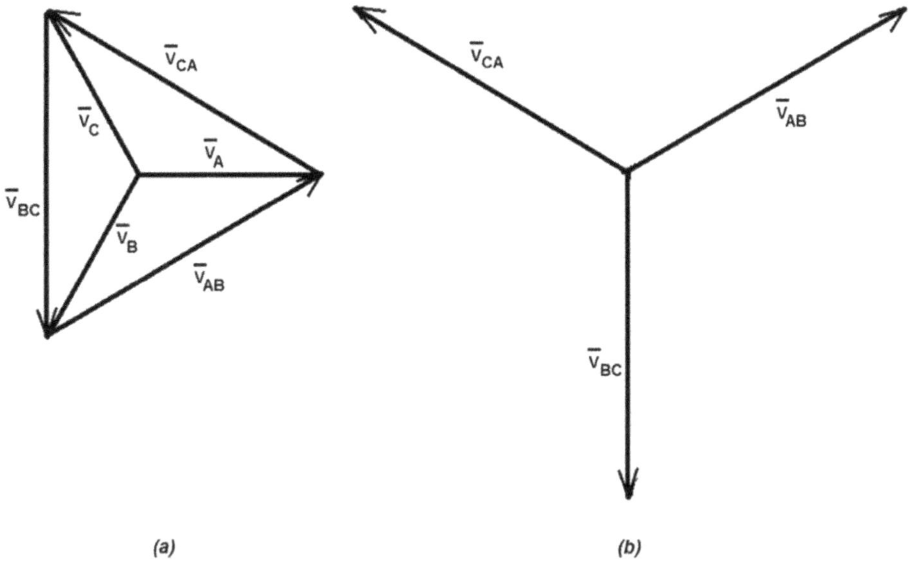

Figure 1.6. Phasor diagram showing (a) line to line voltages as the difference between line to neutral voltages and (b) line to line voltages showing the magnitude and phase angle relationships.

(a) Find the line currents.

From equations (1.19) and (1.20),

$$\overline{I}_A = \frac{7200 \text{ V} \underline{/0^\circ}}{0.25 + j1.0\Omega + 0.6 + j0.9\Omega + 35 + j20\Omega}$$
$$= 146.3 - j89.3 \text{ A} = 171.4 \text{ A} \underline{/-31.4^\circ}$$
$$\overline{I}_B = \frac{7200 \text{ V} \underline{/-120^\circ}}{0.25 + j1.0\Omega + 0.6 + j0.9\Omega + 35 + j20\Omega}$$
$$= -150.5 - j82.0 \text{ A} = 171.4 \text{ A} \underline{/-151.4^\circ}$$
$$\overline{I}_C = \frac{7200 \text{ V} \underline{/120^\circ}}{0.25 + j1.0\Omega + 0.6 + j0.9\Omega + 35 + j20\Omega}$$
$$= 4.2 + j171.3 \text{ A} = 171.4 \text{ A} \underline{/88.6^\circ}$$

Note that the calculations for the B and C phase currents are somewhat repetitive in balanced systems. These can be obtained by simply shifting the A phase result forward and backward by 120°.

(b) Find the load voltage.

The A phase to neutral load voltage can be found as

$$\overline{V}_{A'N} = \overline{Z}_{wye} \cdot \overline{I}_A = (35 + j20\Omega)(171.4\ A \underline{/-31.4^\circ}) = 6909 \text{ V} \underline{/-1.7^\circ}$$

Similarly,

$$\overline{V}_{B'N} = \overline{Z}_{wye} \cdot \overline{I}_B = (35 + j20\Omega)(171.4\ A\ \underline{/-151.4^\circ}) = 6909\ V\ \underline{/-121.7^\circ}$$

$$\overline{V}_{C'N} = \overline{Z}_{wye} \cdot \overline{I}_C = (35 + j20\Omega)(171.4\ A\ \underline{/88.6^\circ}) = 6909\ V\ \underline{/118.3^\circ}$$

Note that the angle differences between source and load voltage is small, as would be typical for this type of power system.

(c) Find the line to line voltages at the load:

The voltage between points A′ and B′ is

$$\overline{V}_{A'B'} = 6909\ V\ \underline{/-1.7^\circ} - 6909\ V\ \underline{/-121.7^\circ} = 11953\ V\ \underline{/28.3^\circ}$$

As expected, the line to line voltage magnitude is $\sqrt{3}$ times the line to neutral voltage magnitude, and the line to line voltage angle leads the A phase line to neutral voltage angle by 30°. Numerical calculation of the B′ to C′ and C′ to A′ voltages shows the same relationships between voltage magnitudes and angles. These are illustrated in figure 1.6.

(d) Find the three phase apparent power and real power at the load.

The three phase apparent power is the sum of the apparent power in each of the three legs of the load. It is important to remember that it is the voltage and current magnitudes that are used in this calculation,

$$\begin{aligned} S_{3\phi} &= S_A + S_B + S_C = V_{A'N}I_A + V_{B'N}I_B + V_{C'N}I_C \\ &= 6909\ V \cdot 171.4\ A + 6909\ V \cdot 171.4\ A + 6909\ V \cdot 171.4\ A \\ &= 3.553 \times 10^6\ VA\ \textit{or}\ 3.553\ MVA \end{aligned}$$

The three phase real power is

$$\begin{aligned} P_{3\phi} &= P_A + P_B + P_C = V_{A'N}I_A\cos(\delta_A - \phi_A) + V_{B'N}I_B\cos(\delta_B - \phi_B) \\ &\quad + V_{C'N}I_C\cos(\delta_C - \phi_C) \\ &= 6909\ V \cdot 171.4\ A \cdot \cos(-1.7^\circ - (-31.4^\circ)) \\ &\quad + 6909\ V \cdot 171.4\ A\cos(-121.7^\circ - (-151.4^\circ)) \\ &\quad + 6909\ V \cdot 171.4\ A\cos(28.3^\circ - 88.6^\circ) \\ &= 3.086 \times 10^6\ W\ \textit{or}\ 3.086\ MW \end{aligned}$$

Note that the argument of each cosine term in this equation is 29.7°. The fact that this angle is the same in each phase is expected due to the symmetry of this circuit.

This example shows that there are many repetitive calculations in this method of solving this circuit. A method for reducing the number of calculations for balanced wye type circuits is given in the next section.

1.3.2 Per phase equivalent circuit

In the previous analysis of balanced three phase wye circuits, once the A phase quantities are known, the B and C phase values follow directly—the circuit in figure 1.5 can be solved by equation (1.19), and the B and C phase currents are found by shifting the A phase current by the appropriate angle. It is therefore sufficient to solve the circuit shown in figure 1.7. This circuit is called the per phase equivalent circuit. It is important to note that this equivalent circuit involves line to neutral voltages rather than line to line voltages. The per phase equivalent circuit can be used to solve for one of the phases of a balanced three phase circuit (a phase is invariably chosen for this). Once this is done, the currents and voltages in the other two phases can be readily determined.

When the per phase equivalent circuit is solved for current and voltages, the apparent, real and reactive power flows for the three phase circuit can be directly determined. As the same power flows in each of the three phases, the total power flowing at some point in the circuit simply three times the per phase power, or:

$$\begin{aligned}\text{apparent power } S_{3\phi} &= 3[V_{\text{LN}}I_{\text{Line}}]\\ \text{real power } P_{3\phi} &= 3[V_{\text{LN}}I_{\text{Line}}][\cos(\delta_{\text{V}} - \phi_{\text{I}})] = S_{3\phi}\cos(\delta_{\text{V}} - \phi_{\text{I}})\\ \text{reactive power } Q_{3\phi} &= 3[V_{\text{LN}}I_{\text{Line}}][\sin(\delta_{\text{V}} - \phi_{\text{I}})] = S_{3\phi}\sin(\delta_{\text{V}} - \phi_{\text{I}})\end{aligned} \tag{1.27}$$

In this equation, the subscript 'LN' for voltage indicates the line to neutral voltage at some point in the circuit, the subscript 'Line' for current indicates that this is the current flowing in the line at that point, and the subscripts 'V' and 'I' on the angles emphasize that these are the angles of the line to neutral voltage and line currents at this point. The subscript 'ϕ' in equation (1.27) is used as a shorthand for 'phase', so '3ϕ' indicates that these are three phase values of power, apparent power, and reactive power.

Note that the power triangle is equally applicable to this three phase circuit as it was to the single phase circuit. In particular, it is worth noting that

$$S_{3\phi}^2 = \sqrt{P_{3\phi}^2 + Q_{3\phi}^2} \tag{1.28}$$

and

$$\text{power factor } pf = \frac{P_{3\phi}}{S_{3\phi}} = \cos(\delta_{\text{V}} - \phi_{\text{I}}) \tag{1.29}$$

Alternate forms of the power equations: There are several alternate forms for the three phase power equations derived in the previous section.

(1) Power in terms of line to line voltages: since the line to line voltages are $\sqrt{3}$ times the line to neutral voltages, equation (1.27) can be rewritten as

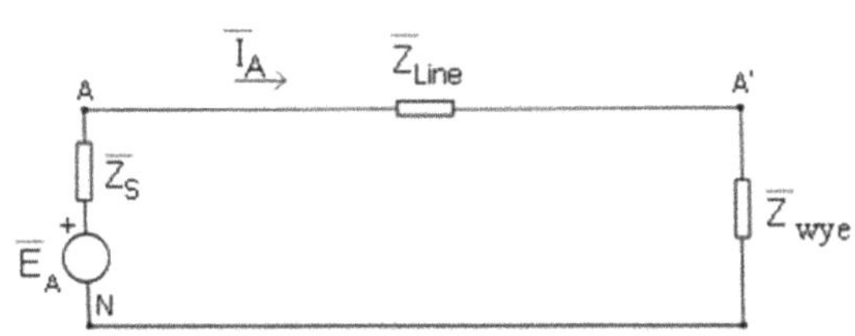

Figure 1.7. Per phase equivalent circuit of the three phase circuit of figure 1.5.

$$\text{real power} P_{3\phi} = \sqrt{3}[V_{L-L}I_{Line}][\cos(\delta_V - \phi_I)]$$
$$\text{apparent power} S_{3\phi} = \sqrt{3}[V_{L-L}I_{Line}] \tag{1.30}$$
$$\text{reactive power} Q_{3\phi} = \sqrt{3}[V_{L-L}I_{Line}][\sin(\delta_V - \phi_I)]$$

It must be noted in these equations that the voltage angle δ_V is the angle of the line to neutral voltage, not the line to line voltage. For this reason, some feel that it is awkward to use the equations in this form.

(2) Complex power: due to the geometrical relationship of the power triangle, a complex three phase power $\overline{S}_{3\phi}$ can be defined,

$$\overline{S}_{3\phi} = P_{3\phi} + jQ_{3\phi} = 3\overline{V}_{LN}\overline{I}^{*}_{Line} \tag{1.31}$$

where $(\cdot)^{*}$ refers to the complex conjugate operator. The magnitude of the complex apparent power $\overline{S}_{3\phi}$ is the magnitude to the apparent power $S_{3\phi}$, and the angle of the complex power $\overline{S}_{3\phi}$ is the angle of the power triangle, shown in figure 1.2. Note that this equation refers to the line to neutral voltage and line current in the same phase, and assumes balanced three phase conditions.

1.3.3 Delta connected loads

Power system loads and sources can be connected in delta rather than in wye. A delta connected load is shown in figure 1.8. The load is balanced, and is fed by a wye connected source through a transmission line. The delta connected load has no neutral point, and so there is no connection between the load and the source neutral point.

In this form, this circuit is more difficult to solve than the circuit of figure 1.5. It could be solved directly by writing three loop equations or three node equations. Neither of these methods take advantage of the symmetry of this balanced circuit, and require the solution of three simultaneous equations.

It is preferable to solve this circuit by making a delta to wye conversion of the load. As the three delta impedances are the same, the equivalent wye impedances are

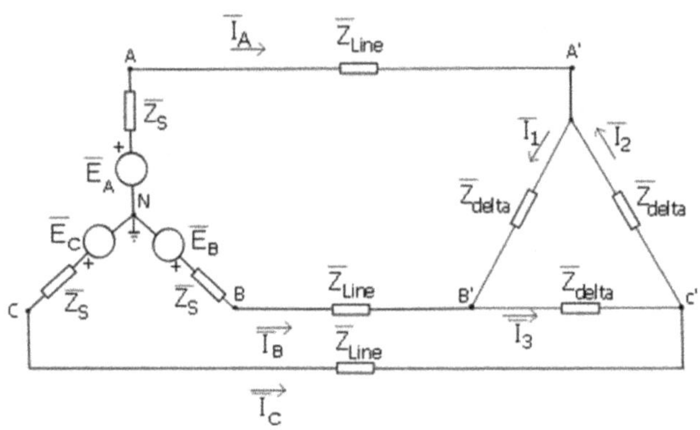

Figure 1.8. Three phase circuit serving a delta connected load.

$$\overline{Z}_{\text{wye}} = \frac{\overline{Z}_{\text{delta}}}{3} \tag{1.32}$$

After making this substitution, the circuit is now equivalent to the wye load circuit of figure 1.5. A per phase equivalent circuit can then be created, and the circuit can be solved through the methods of the previous section. This solution will give the line currents and the line to neutral and line to line voltages. The currents in each leg of the delta connected load can then be found from the load voltage and delta impedance,

$$\overline{I}_1 = \frac{\overline{V}_{A'B'}}{\overline{Z}_{\text{delta}}} \tag{1.33}$$

It can be shown that these currents in the load are smaller than the line currents by a factor of $\sqrt{3}$, and are shifted by 30°.

Mathematically, from figure 1.8, the phase A line current is

$$\overline{I}_A = \overline{I}_1 - \overline{I}_2 \tag{1.34}$$

For a balanced system, if $\overline{I}_1 = I_d\underline{/\theta_d}$, then $\overline{I}_2 = I_d\underline{/\theta_d - 120°}$

From equation (1.34),

$$\overline{I}_A = I_d\underline{/\theta_d}(1\underline{/0°} - 1\underline{/-120°}) = \sqrt{3}\,I_d\underline{/\theta_d + 30°} \tag{1.35}$$

In balanced three phase systems, delta connected loads or sources should be analysed by replacing the delta circuit with an equivalent wye circuit. This circuit should then be solved by the per phase equivalent circuit. Once the per phase circuit is solved, circuit voltage, currents, and power can be readily found.

Example 1.5. In the balanced three phase circuit shown in figure 1.8, the A phase source voltage is $\overline{E}_A = 277\text{V}\underline{/0°}$. The source impedance is $\overline{Z}_s = 0.02 + j0.05\Omega$, the line impedance is $\overline{Z}_{\text{Line}} = 0.04 + j0.08\Omega$, and the delta connected load impedance is $\overline{Z}_{\text{delta}} = 12 - j1.5\Omega$.

(a) Draw the per phase equivalent circuit for this case.

The per phase diagram is based on wye connected sources and loads. The wye equivalent impedance for the delta load is

$$Z_{\text{wye}} = \frac{Z_{\text{delta}}}{3} = 4.0 - j0.5\Omega$$

The resulting per phase equivalent circuit is:

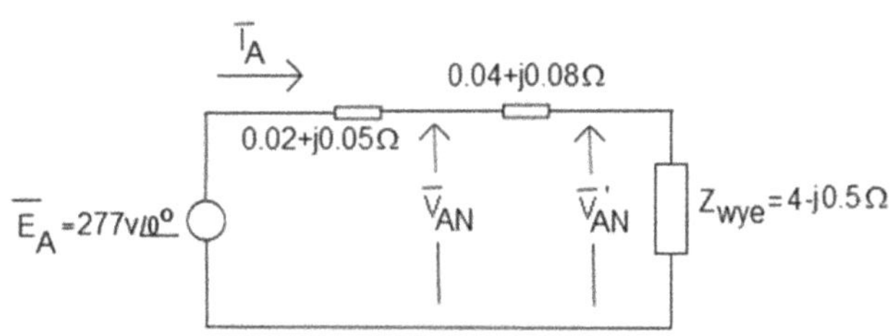

(b) Find the A phase load current and voltage for this circuit:
From the per phase equivalent circuit, the current $\bar{I}_A$ is

$$\bar{I}_A = \frac{277\text{V}\underline{/0^\circ}}{(0.02 + j0.05)\Omega + (0.04 + j0.08)\Omega + (4 - j0.5)\Omega}$$

Solving,

$$\bar{I}_A = 67.66 + j6.17\text{A} = 67.94\text{A}\underline{/+5.2^\circ}$$

The load voltage is then

$$\bar{V}_{A'N} = \bar{Z}_{wye}\bar{I}_A = 273.9\text{V}\underline{/-1.9^\circ}$$

(c) Find the magnitude of the line to line voltages at the load:

$$V_{LL} = \sqrt{3}\,V_{LN} = 1.732 \cdot 273.9 = 474.4\text{V}$$

(Note: since in the balanced case all three line to line voltages are equal, and all three line to neutral voltages are equal, generic subscripts are often used for these values.)

(d) Find the magnitude of the current in the delta load:

From equation (1.35), $I_{delta} = \frac{1}{\sqrt{3}} I_{Line} = \frac{67.94\text{A}}{1.732} = 39.23\text{A}$

(e) Find the real power drawn by the load:
From equation (1.27),

$$P_{Load} = 3V_{A'N}I_A\cos(\delta_V - \phi_I) = 3 \cdot 273.9\text{V} \cdot 67.94\text{A} \cdot \cos(-1.9^\circ - 5.2^\circ)$$
$$= 55.83\text{KW}$$

1.4 Summary

In today's society, we use significant amounts of energy in our everyday lives. This energy comes from a variety of different sources, and is used in a wide range of applications. Electric energy is arguably the most convenient and flexible source of energy for usage to meet our residential, commercial, industrial, and perhaps transportation needs. Our primary energy sources are converted to electrical energy, transported to the consumer by transmission and distribution lines, and then converted from electrical energy into some other form, such as motion, light, heat, cooling, or for digital devices.

Reference

[1] Annual Energy Outlook 2013 With Projections to 2040 April 2013 US Energy Information Administration, Office of Integrated and International Energy Analysis (Washington, DC: US Department of Energy)

Questions

1. Download the most recent Annual Energy Outlook from the US Energy Information Administration. Use data from this report to update table 1.5, on the energy sources for electric power. From this result, discuss the trends in renewable energy generation.
2. Describe the difference between RMS values and peak values for alternating voltages and currents. Why are RMS values generally used for voltage and current in electric power systems?

Problems

1. A 2.5 MW wind turbine will typically operate at an average level of 28% of its rated capacity over the course of the year.
    a. Determine the megawatt-hours (MWh) that would be delivered by this turbine in a year.
    b. In 2015, the electric power generated in coal plants was about 18.5 EJ. Determine the number of these wind turbines that would be needed to reduce coal consumption by 1%.
2. An electric load is drawing power at the rate shown below. Determine the energy delivered to the load over these 6 h.

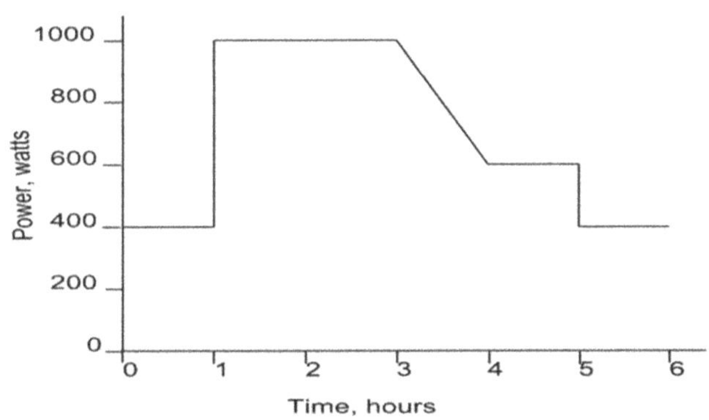

3. A single phase AC circuit is shown below.
    a. Find the phasor current $\overline{I}_s$ and voltage $\overline{V}_X$.
    b. Find the apparent power S_x flowing at point x in the circuit.
    c. Also, find the watts and VARs flowing at X.
    d. Find the power factor at point X.

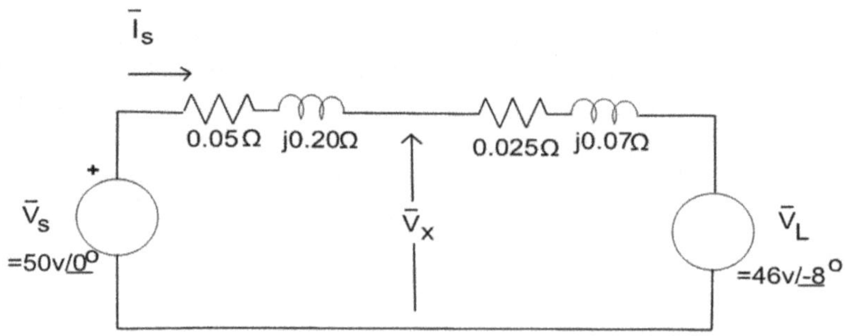

4. Using trigonometric substitutions, find the value of the line to line voltage phasor $\overline{V}_{AB} = \overline{V}_A - \overline{V}_B$ when $\overline{V}_A = V_t\underline{/0^\circ}$ and $\overline{V}_B = V_t\underline{/-120^\circ}$
5. A balance set of three phase currents flows into a grounded wye load. The currents are $\overline{I}_A = 20\text{A}\,\underline{/0^\circ}$, $\overline{I}_B = 20\text{A}\,\underline{/-120^\circ}$, $\overline{I}_C = 20\text{A}\,\underline{/+120^\circ}$. Find the current flowing from the neutral point to ground.
6. A balanced three phase system has a wye source and wye load, as shown below. The source voltages are balanced, and have a line to neutral magnitude of 277 V. The system impedances are $\overline{Z}_s = 0.05 + j0.20\Omega$, $\overline{Z}_{Line} = 0.15 + j0.50\Omega$, and $\overline{Z}_{wye} = 12 + j2\Omega$.
   a. Draw the per phase equivalent circuit for this system.
   b. Determine the A phase line current flow.
   c. Find the load voltage $\overline{V}_{A'}$.
   d. Find the three phase volt-amps and real power flow at the load.

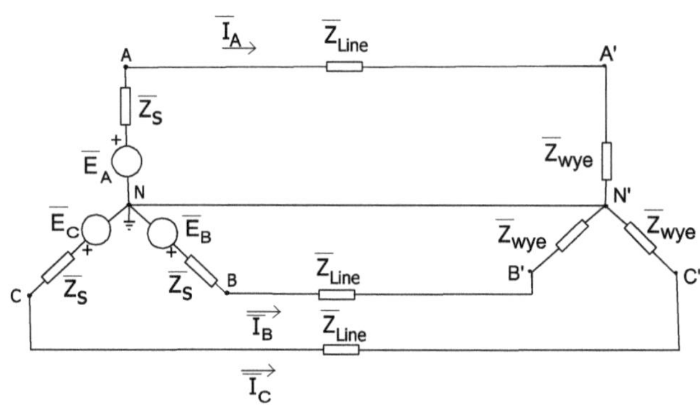

7. For the wye source wye load circuit shown above, the load voltage is balanced, and has a line to line magnitude of 4160 V. The three phase apparent power is 450 kVA. The load has a power factor of 0.85 leading.
   a. Draw the per phase equivalent circuit for this system.
   b. Find the magnitude of the phase currents.
   c. Find the angle of the A phase current relative to the A phase load voltage.
   d. Determine the source voltage $\overline{E}_A$.

8. A balanced three phase system is shown below. The source voltages are balanced, and the A phase source voltage is $\overline{E}_A = 7600\text{V}\,\underline{/0°}$. The system impedances are $\overline{Z}_s = 0.2 + j1.2\Omega$, $\overline{Z}_{Line} = 0.6 + j3.0\Omega$, and $\overline{Z}_{delta} = 42 + j12\Omega$.
   a. Draw the per phase equivalent circuit for this system.
   b. Find the line current in each phase.
   c. Find the A phase voltage A′ between the load and ground.
   d. Find the magnitude to the line to line voltage at the load.
   e. Find the three phase volt-amps, watts, and VARs at the load.

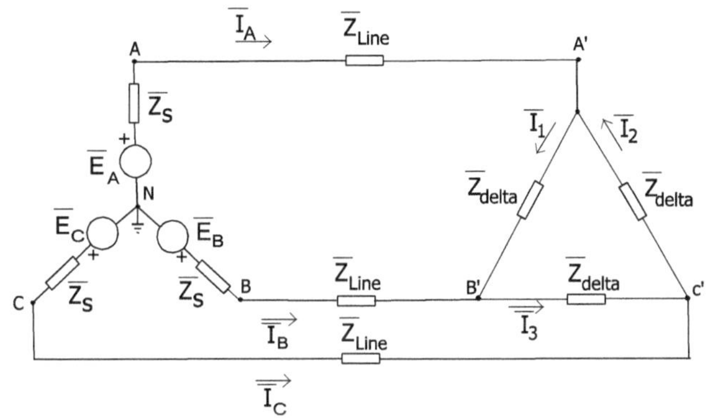

IOP Publishing

Electromechanical Machinery Theory and Performance

Thomas Ortmeyer

Chapter 2

Magnetic circuits

The ferromagnetic properties of iron and other metals are used to provide an important set of electrical equipment. In particular, the transformers and rotating machines that are the subject of this text are made practical by these properties.

Typical soft ferromagnetic materials exhibit relative permeability in the range of several thousand. As a result, the magnetic flux in these machines will reside primarily in the ferromagnetic cores of these devices. For an initial analysis, the magnetic fields in these ferromagnetic cores can be considered to be one dimensional, which considerably simplifies the analysis. This type of analysis is generally referred to as magnetic circuit analysis. In many cases, the initial designs of these devices are done with magnetic circuit analysis, and then refined with two or three dimensional finite element analysis.

2.1 Rectangular core magnetic circuit

Figure 2.1 shows a rectangular magnetic core with a single coil of wire. The flow of current in the wire creates a magnetic field. This relationship is defined by Ampere's law,

$$\oint \boldsymbol{H} \cdot d\boldsymbol{\ell} = i_{\text{enc}} \tag{2.1}$$

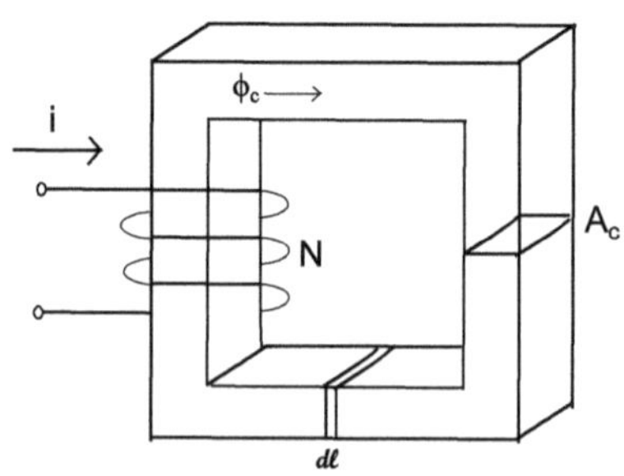

Figure 2.1. Rectangular magnetic core with a single coil.

doi:10.1088/978-0-7503-1662-0ch2

In this equation, $\boldsymbol{H}$ is the vector magnetic field, and $d\boldsymbol{\ell}$ is an incremental length vector. This equation states that the integral of the magnetic field around a closed path $\boldsymbol{\ell}$ equals the total amount of current that passes through a window created by that closed path, labeled i_{enc}. This equation is generally true at low frequencies. Ampere's law in this form is general but can be difficult to solve. In many cases, the only method of solution is with the numerical solution of this partial differential equation over multiple dimensions. In some cases, however, approximate solutions can be obtained by making simplifying assumptions. Magnetic circuits involving high permeability iron fall into this category.

In figure 2.1, the magnetic circuit is a rectangular magnetic core with an N turn coil wrapped around the core. The high permeability of iron means that the majority of the magnetic flux will exist in the core. While a small amount of flux will exist outside of the core, it can be neglected at this point.

The path $\boldsymbol{\ell}$ will be taken along the length of the core, as shown in the figure. The iron core has the same characteristics all around its length (with a brief exception at each corner). As a result, the magnetic field will be constant at any point in the core. The line integral then becomes a simple product of the core field intensity and the length of the core.

$$H_c \ell_c = N i_{\text{coil}} \tag{2.2}$$

$\boldsymbol{\ell}_c$ is the effective core length, the sum of the effective lengths of the two vertical legs and the two horizontal legs. N is the number of turns of the coil, and i_{coil} is the current in this coil. The product of turns times current is the value i_{enc} of equation (2.1) that passes through the window created by the iron core. This product is stated in units of amp-turns.

The amp-turns provided by a coil are often referred to as the magnetomotive force, MMF, of the magnetic circuit. This MMF creates a magnetic field in the iron core, in a similar fashion to an electromotive force (voltage source) in an electric circuit. In MKS units, the magnetic field is expressed in units of amp-turns per meter, and the core length is stated in meters.

To get an effective measure of the mean core length, each leg is measured from the center of the connecting leg. Consider the core of figure 2.2. The width of both vertical legs and both horizontal legs is W.

The effective horizontal length of this core is

$$\ell_h = L_1 - \frac{1}{2}W - \frac{1}{2}W$$

Also, the effective vertical length is

$$\ell_v = L_2 - \frac{1}{2}W - \frac{1}{2}W$$

The total effective length of the core is the sum of the two horizontal lengths plus the two vertical lengths,

$$\ell_c = 2\ell_h + 2\ell_v$$

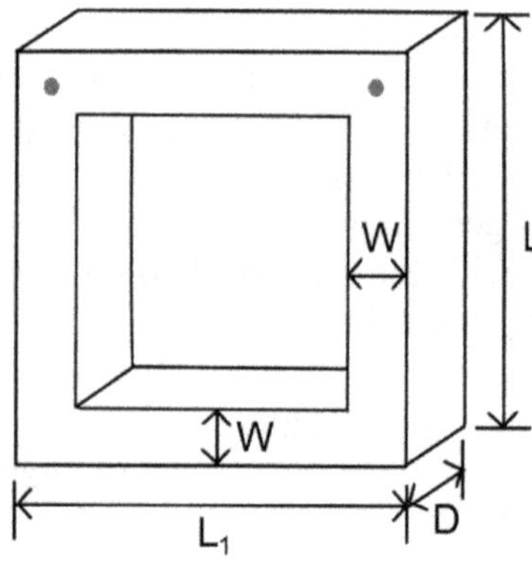

Figure 2.2. Length, width, and depth definitions for the simple magnetic core. The distance between dots on top is the effective horizontal length.

The adjustments in length are made to approximate the average path that flux takes around the core.

Note that in this case, the width W of the core is identical in all legs of the rectangle, as is the depth D of the core. While not unusual, this is not always the case.

Flux density B_c is the product of the magnetic field intensity and the permeability of the iron core,

$$B_c = \mu H_c = \mu_R \mu_o H_c \tag{2.3}$$

Flux density is expressed in units of tesla, which is equivalent to webers per meter squared. The permeability of the core is expressed as the product of the relative permeability of the core material μ_R and permeability of free air μ_o, where

$$\mu_o = 4\pi \times 10^{-7} \text{ henries per meter} \tag{2.4}$$

Any closed path within the iron core will enclose the same number of amp-turns. This will lead to a uniform flux density over the core cross section. Therefore, the core flux will equal the product of the flux density and the cross sectional area of the core,

$$\phi_c = B_c A_c \tag{2.5}$$

Here, the units of flux ϕ_c are webers. A_c is the cross sectional area of the core, the product of the core width and core depth, with units of square meters. The direction of magnetic flux flow in the circuit is determined by the right hand rule. Wrap the fingers of your right hand around the coil in the direction of current flow. Extend your thumb out perpendicular to the current flow. Your thumb is pointing in the direction of the flow of flux.

Example 2.1. A rectangular iron core is shown in figure 2.3. The core has a 150 turn coil that is carrying 0.5 A. The relative permeability of the iron is $\mu_R = 2500$.

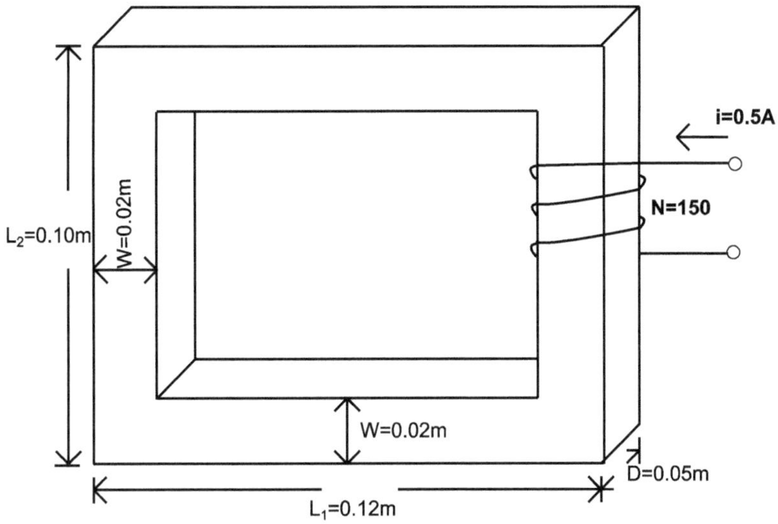

Figure 2.3. Rectangular iron core with 150 turn coil.

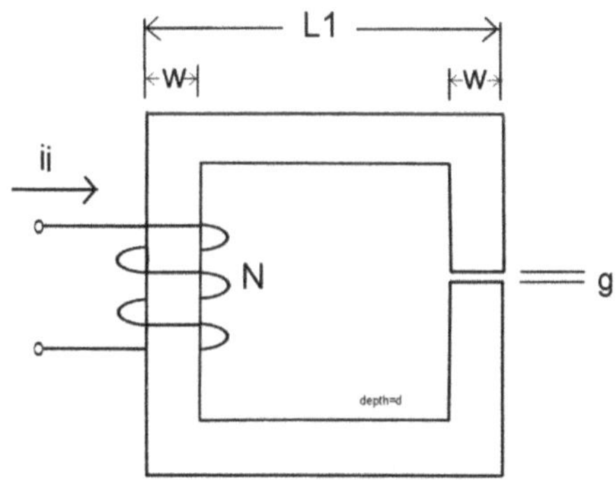

Figure 2.4. Simple magnetic circuit with air gap.

(a) Find the effective core length and cross sectional area:

$$\ell_C = 2(L_1 - W) + 2(L_2 - W) = 2 \cdot 0.10\text{ m} + 2 \cdot 0.08\text{ m} = 0.36\text{ m}$$

$$A_c = W \cdot D = 0.05\text{ m} \cdot 0.02\text{ m} = 0.001\text{ m}^2$$

(b) Find the MMF for this circuit

$$\mathscr{F} = Ni = 75\text{ A} - \text{t}$$

(c) Find the magnetic field present in the iron core

$$\mathscr{H}_c = \frac{\mathscr{F}}{\ell_c} = \frac{75\text{ A} - \text{t}}{0.36\text{ m}} = 208.3\text{ amp} - \text{turns/meter}$$

(d) Find the flux density in the core

$$\mathscr{B}_c = \mu_R \mu_o \mathscr{H}_c = (2500) \cdot (4\pi \times 10^{-7}) \cdot 208.3\ (\text{A} - \text{T})/\text{m} = 0.6545 \text{ tesla}$$

(e) Find the total core flux

$$\phi_C = A_c \mathscr{B}_c = 0.001 \text{ m}^2 \cdot 0.6545\text{T} = 6.545 \times 10^{-4} \text{ webers or } 0.6545 \text{ milliwebers}$$

2.2 Magnetic circuits with small air gap

In this equation, the symbol $\mathscr{F}$ is used to refer to the magnetomotive force applied by the coil current. The magnetic field in the core H_c is constant through the effective core length ℓ_c. In the core, the magnetic field H_g is constant in the air gap through the gap length g.

If the permeability of the core and the gap are known, this equation can be rewritten as

$$\mathscr{F} = \frac{B_c}{\mu_o \mu_R} \ell_c + \frac{B_g}{\mu_o} g \tag{2.6}$$

The total flux will be the same everywhere in this uniform magnetic core. The flux will be largely contained in the iron core where it exists, due to the high permeability of iron. In the gap region, the flux will tend to spread out, as shown in figure 2.5. This is termed as the fringing effect. It can be determined most accurately by solving Ampere's equation in partial differential equation form using the finite element method. For our purposes, however, the fringing effect can be approximated by modeling the air gap as having a larger cross sectional area than the iron core. For short gap lengths, the fringing affect can increase the apparent cross sectional area by 5–10%, depending on the gap length relative to the core cross sectional area,

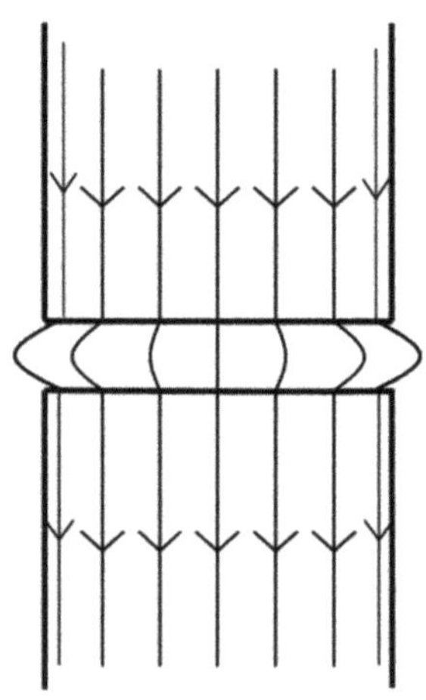

Figure 2.5. Close up of air gap showing fringing of the magnetic flux.

$$A_g = A_c(1 + \eta) \tag{2.7}$$

where η = (% increase in effective gap size)/100

Using this relationship, equation (2.6) can be rewritten as

$$\mathscr{F} = \left(\frac{L_c}{\mu_o \mu_R A_c} + \frac{g}{\mu_o (1 + \eta) A_c} \right) \phi_c \tag{2.8}$$

Note that the total flux ϕ_c is the same in the core and in the gap, while the flux density differs in the core and gap due to the fringing effect.

Equation (2.8) shows the amp-turns applied by the coil current creates core flux ϕ_c. The core flux created is a function the properties of the core and of the air gap. In this magnetic circuit, the MMF is analogous to the electromotive force (voltage source) in an electric circuit. The flux ϕ_c is analogous to the current in an electric circuit, and the terms describing the core and gap are analogous to circuit resistance. The term 'reluctance' has been defined to describe these terms. The reluctance of the core is

$$\mathfrak{R}_c = \frac{L_c}{\mu_o \mu_R A_c} \tag{2.9}$$

Also, the reluctance of the air gap is

$$\mathfrak{R}_g = \frac{g}{\mu_o (1 + \eta) A_c} \tag{2.10}$$

With these terms defined, the core flux can be found to be

$$\phi_c = \frac{\mathscr{F}}{\mathfrak{R}_c + \mathfrak{R}_g} \tag{2.11}$$

The form of this equation is the same as for an electrical circuit with two resistors connected in series. This magnetic circuit can be represented by its electrical analog, where the MMF $\mathscr{F}$ looks like a voltage source, the flux is equivalent to the current, and the core and gap reluctances are represented by resistances. The corresponding electrical analog of this magnetic circuit is shown in figure 2.6.

Magnetic circuit analysis using the reluctance method can be used to analyse complicated magnetic circuits in the same fashion that circuit theory is used to analyse complex electrical circuits.

Example 2.2. Repeat example 2.1, but use the magnetic circuit reluctance to calculate the core flux.

(a) Magnetic circuit reluctance

$$\mathscr{R}_C = \frac{L_C}{\mu_R \mu_o A_c} = \frac{0.36 \text{ m}}{(2500)(4\pi \times 10^{-7})(0.001 \text{ m}^2)}$$

$$= 114\,600 \text{ amps/weber or } 114.6 \text{ kiloamps/weber}$$

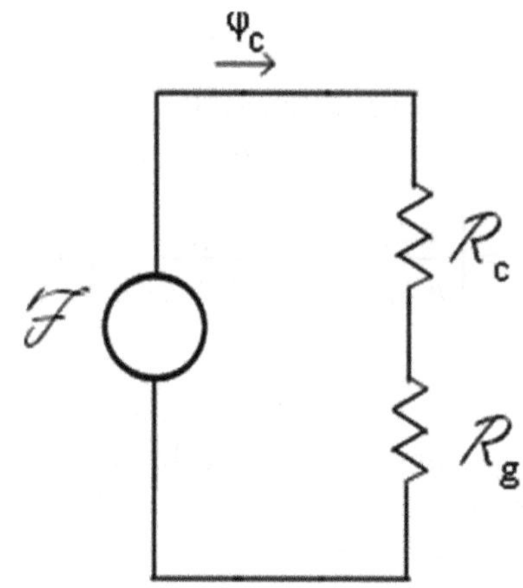

Figure 2.6. Electrical analog of the magnetic circuit of figure 2.4.

(b) Core flux. With no air gap, $\mathfrak{R}_g = 0$, so

$$\phi_c = \frac{\mathscr{F}}{\mathfrak{R}_c} = \frac{75A - t}{114\,600 \text{ amps/weber}} = 0.000\,655 \text{ webers or } 0.655 \text{ milliwebers}$$

Example 2.3. Repeat example 2.2, but include an air gap of $g = 0.0025$ m in the iron core. The fringing factor for this case is $\eta = 0.04$.

(a) Neglect the impact of the air gap on the effective core length L_c. The air gap reluctance is

$$\mathfrak{R}_g = \frac{g}{\mu_o A_c(1 + \eta)} = \frac{0.0025 \text{ m}}{(4\pi \times 10^{-7})(0.001 \text{ m}^2)(1.04)} = 1.912 \times 10^6 \text{ amps/weber}$$

$$\phi_c = \frac{\mathscr{F}}{\mathfrak{R}_c + \mathfrak{R}_g} = \frac{75\,A - t}{0.1146 \times 10^6 \text{ amps/weber} + 1.912 \times 10^6 \text{ amps/weber}}$$
$$= 0.370 \cdot 10^{-6} \text{ webers or } 0.370 \text{ microwebers}$$

These examples show that a very small air gap will have a major impact on the amount of flux generated by a given current. The air gap does have other desirable properties, however, and is commonly used in some applications.

2.3 Three leg core

Magnetic circuits come in a variety of forms. Another common form is the three legged core, shown in figure 2.7. For the moment, we will consider a single coil on this magnetic circuit.

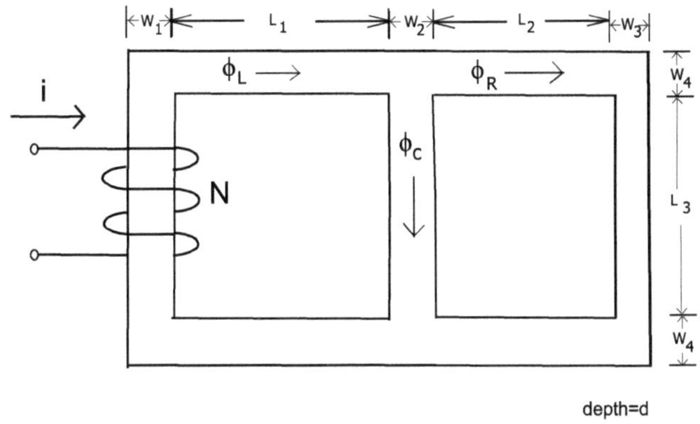

Figure 2.7. Three legged core with one coil.

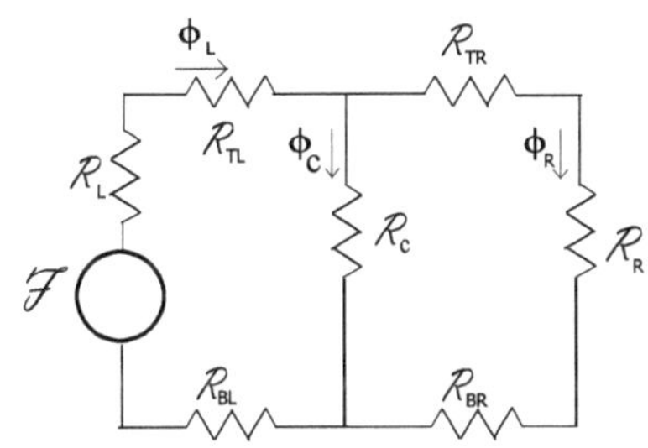

Figure 2.8. Electrical equivalent circuit for the magnetic circuit of figure 2.7.

The magnetic core has seven segments—three vertical legs, two horizontal connectors on the top, and another two on the bottom. Each segment has length, cross section, and permeability. The reluctance of each segment is therefore

$$\mathscr{R}_X = \frac{\ell_X}{\mu_R \mu_o A_X} \tag{2.12}$$

The MMF $\mathscr{F} = Ni$ creates flux in the core. The reluctance of each leg of the core can be calculated, and the flux found by using an electrical equivalent circuit. The equivalent circuit for this magnetic circuit is shown in figure 2.8.

This circuit can be solved to find the fluxes ϕ_L, ϕ_C and ϕ_R, in the left, center and right legs, respectively. This method can be used to solve complex magnetic circuits where the relative permeability μ_R is known. The effective length of the top left leg of the core is

$$\ell_{TL} = L_1 + \frac{1}{2}(W_1 + W_2) \tag{2.13}$$

The depth of the core is the same for the whole core. The cross sectional area for each segment is the product of the leg width times the leg depth.

The electrical analog for the magnetic circuit of figure 2.7 is shown in figure 2.8. This circuit is laid out so the reluctances in the analog circuit have the same layout as the legs of the magnetic core.

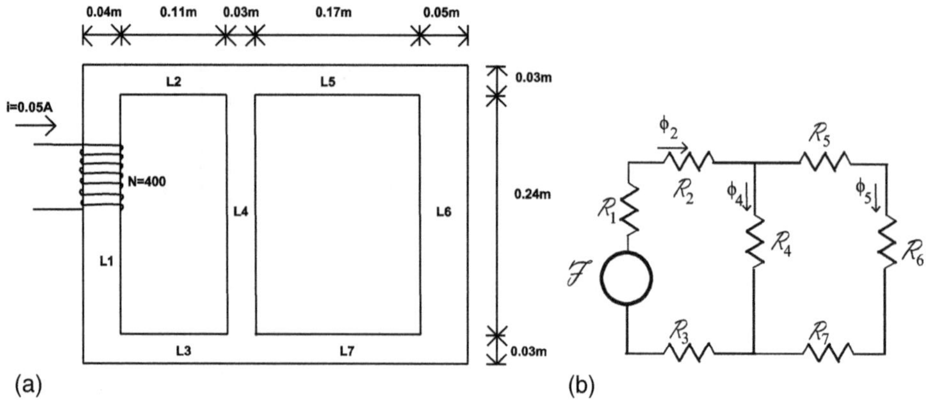

Figure 2.9. (a) Magnetic core for example 2.4 and (b) electric analog circuit for example 2.4.

Example 2.4. Determine the core flux in the magnetic core shown in figure 2.9. The relative permeability of the core is $\mu_R = 1800$, and the core depth $d = 0.04$m.

Solution. The iron core is divided into seven legs. These are noted as L1 through L7 in figure 2.9. The reluctance for each leg is:

$$\mathcal{R}_1 = \frac{\ell_1}{\mu_R \mu_o A_{c1}} = \frac{(0.24 + 0.03)\ \text{m}}{1800(4\pi \times 10^{-7})(0.04\ \text{m} \cdot 0.04\ \text{m})} = 74\,600\ \text{amps/weber}$$

$$\mathcal{R}_2 = \frac{\ell_2}{\mu_R \mu_o A_{c2}} = \frac{(0.11 + 0.02 + 0.025)\ \text{m}}{1800(4\pi \times 10^{-7})(0.03\ \text{m} \cdot 0.04\ \text{m})} = 57\,100\ \text{amps/weber}$$

$$\mathcal{R}_3 = \mathcal{R}_2 = 57\,100\ \text{amps/weber}$$

$$\mathcal{R}_4 = \frac{\ell_4}{\mu_R \mu_o A_{c4}} = \frac{(0.24 + 0.03)\ \text{m}}{1800(4\pi \times 10^{-7})(0.03\ \text{m} \cdot 0.04\ \text{m})} = 95\,790\ \text{amps/weber}$$

$$\mathcal{R}_5 = \frac{\ell_5}{\mu_R \mu_o A_{c5}} = \frac{(0.17 + 0.015 + 0.025)\ \text{m}}{1800(4\pi \times 10^{-7})(0.03\ \text{m} \cdot 0.04\ \text{m})} = 77\,370\ \text{amps/weber}$$

$$\mathcal{R}_6 = \frac{\ell_6}{\mu_R \mu_o A_{c6}} = \frac{(0.24 + 0.015 + 0.015)\ \text{m}}{1800(4\pi \times 10^{-7})(0.05\ \text{m} \cdot 0.04\ \text{m})} = 59\,680\ \text{amps/weber}$$

$$\mathcal{R}_7 = \mathcal{R}_5 = 77\,370\ \text{amps/weber}$$

The resulting electrical analog of this magnetic circuit is shown in figure 2.9. Note that the core flux is generated by the MMF $\mathscr{F}$, and flows up L1, and over L2. It then splits, part of it going through L4 and the rest going through L5, L6 and L7. It then returns through L3 to complete the loop.

The effective circuit reluctance is

$$\mathscr{R}_{\text{effective}} = \mathscr{R}_1 + \mathscr{R}_2 + \mathscr{R}_3 + \mathscr{R}_4 // (\mathscr{R}_5 + \mathscr{R}_6 + \mathscr{R}_7)$$

Numerically, this is

$$\mathscr{R}_{\text{effective}} = 74\,600 + 57\,100 + 57\,100 + \frac{95\,790(77\,370 + 59\,680 + 77\,370)}{95\,790 + (77\,370 + 59\,680 + 77\,370)}$$

$$= 272\,510 \text{ amps/weber}$$

The core flux in leg 1 is then

$$\phi_1 = \frac{\mathscr{F}}{\mathscr{R}_{\text{effective}}} = \frac{20 \text{ A} - \text{t}}{272\,510 \dfrac{\text{amps}}{\text{weber}}} = 73.39 \times 10^{-6} \text{ webers} = 73.39 \text{ microwebers}$$

This same flux flows through legs 2 and 3. The flux in leg 4 can be determined by the current divider rule of electric circuits,

$$\phi_4 = \phi_1 \frac{\mathscr{R}_5 + \mathscr{R}_6 + \mathscr{R}_7}{\mathscr{R}_4 + \mathscr{R}_5 + \mathscr{R}_6 + \mathscr{R}_7} = 43.40 \text{ microwebers}$$

$$\phi_5 = \phi_6 = \phi_7 = \phi_1 - \phi_4 = 30.00 \text{ microwebers}$$

The MMF across leg L4 will be

$$\mathscr{F}_4 = \mathscr{F} - \phi_1(\mathscr{R}_1 + \mathscr{R}_2 + \mathscr{R}_3)$$

$$= (20 \text{ amp} - \text{turns}) - 73.39 \times 10^{-6}(206\,300 \text{ amps/weber}) = 4.86 \text{ A-t}$$

This shows that approximately 15 of the 20 A-t of MMF are required to create the flux in legs 1–3, and the remaining five are needed to create the magnetic flux in legs 4–7.

2.4 Properties of magnetic materials

Saturation and hysteresis

In the previous sections, it has been assumed that the relative permeability of the core material is known and constant. In practice, the relative permeability of ferromagnetic materials varies with flux level.

Ferromagnetic materials exist due to their electron spin characteristics at the atomic level. The set of ferromagnetic elements have electron spins that tend to align with each other when a magnetic field is applied. In small domains that are randomly oriented, there is no net magnetic effect. As a magnetomotive force applies a magnetic field to the material, the domains will begin to align. This creates significantly more magnetic flux per amp-turn than would be the case with non-ferromagnetic materials. This effect continues until the domains are well aligned. At that point, further increase in magnetic field will not result in further alignment, and the rate of increase in flux will decline. This condition is known as saturation. In terms of relative permeability, the relative permeability starts at a low value, increases to as high as several thousand in the range of practical application, and then drops again at high fields, to a value of one in cases of extreme saturation.

The other property of ferromagnetic materials is that they can have a property of remanence, retaining their magnetism when the external magnetic field is removed. Certain classes of ferromagnetic materials have high remanence. Permanent magnets are generally made from these high remanence materials. Permanent magnets have a number of practical applications, including being used to create the magnetic fields in permanent magnet synchronous machines and brushless DC machines.

Materials with strong permanent magnetic properties are commonly referred to as hard magnetic materials. Materials with soft magnetic properties have characteristics suitable for use in alternative magnetic fields.

There are three curves that are used to characterize magnetic materials used in electromagnetic machines. A DC magnetization curve is shown in figure 2.10. This curve starts with the material having 0 flux density when the applied field is 0 amp-turns. As the field is increased, the flux density initially increases slowly, then more quickly and nearly linearly. As the flux density increases above approximately 1 T, the material starts to enter saturation and the rate of flux density increase slows

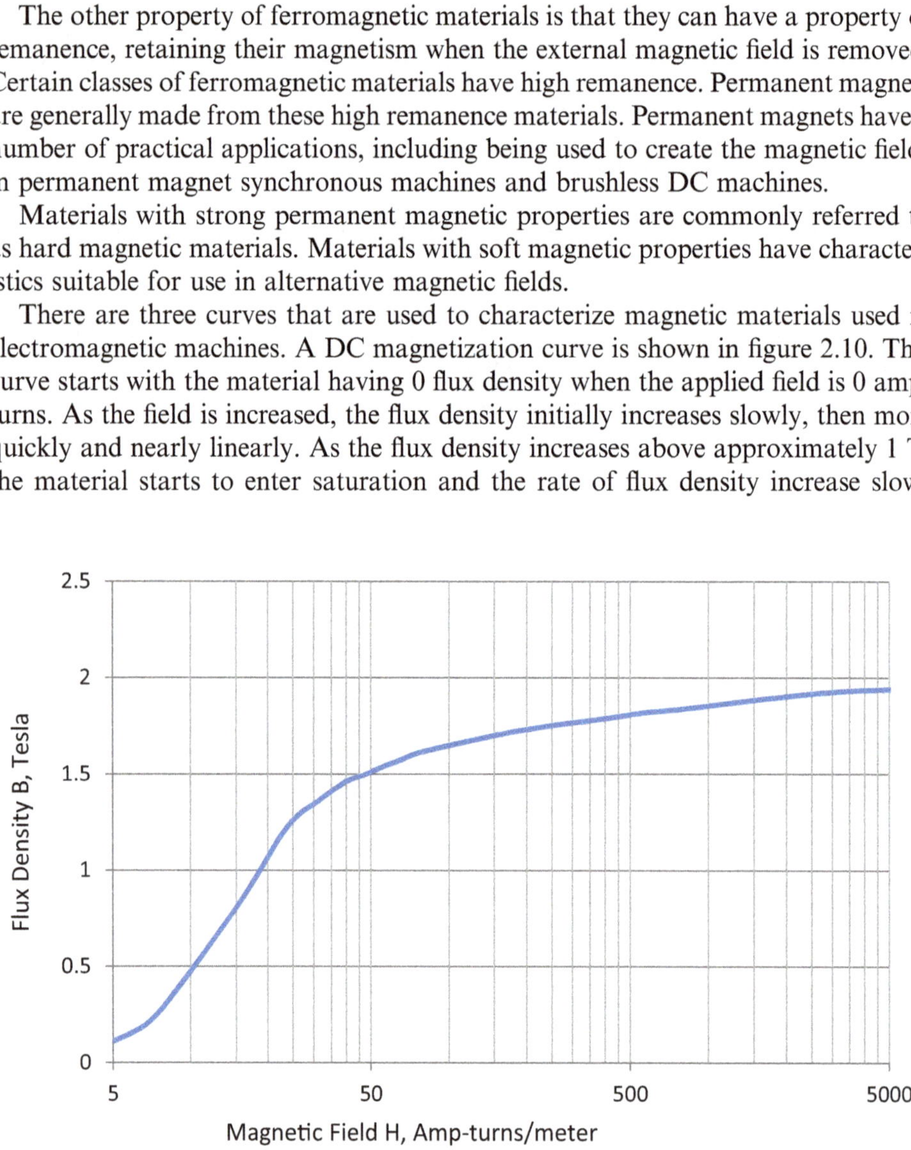

Figure 2.10. DC magnetization curve

markedly. In full saturation, the incremental permeability of the iron drops to the permeability in free space.

Figure 2.11 shows a steady state hysteresis curve for a typical 'soft' magnetic material. This double valued curve shows the large scale impact of the realignment of the magnetic domains at the atomic level. As a steady alternating magnetic field H_c is applied to this material, the flux density B_c is created. The variation of flux density occurs in the direction shown by the arrows on the diagram. In an AC field, the domains continually re-orient themselves. The two different levels of flux density at a given field are a result of the energy that it takes to accomplish this reorientation. As the curve shows, the orientation of the magnetic flux will change directions as the magnetizing field changes.

Steady state alternating magnetic excitation at lower levels will results in smaller curves, with different saturation and hysteresis characteristics.

An interesting point on this curve occurs at $H_c = 0$. This is the residual magnetism left in the material when the magnetic field is brought to zero and kept there. This is often referred to as the remanence of the magnetic material. The remanence in this soft magnetic materials is relatively low, and the hysteresis window is relatively narrow, making this material appropriate for use in AC applications, where the fields are constantly changing.

The final curve that is commonly used for soft magnetic materials is the AC saturation curve. An example AC saturation curve is shown in figure 2.12. This curve shows the B–H relationship over the range of AC excitation. The curve is a plot of the peak value of flux density versus the peak value of the applied magnetic field over the range of sinusoidal excitation at a given frequency. It ignores the hysteresis of the material, and can be used in machine design to determine the magnetic field needed to create a desired level of flux density.

The designer of an AC iron core inductor would generally start with the AC saturation curve for the core material along with an idea of the core geometry. If the design of an AC inductor on the rectangular core shown in figure 2.1, for a peak flux level of ϕ_{design} and for a current of i_{max}, it would be logical to select a peak flux

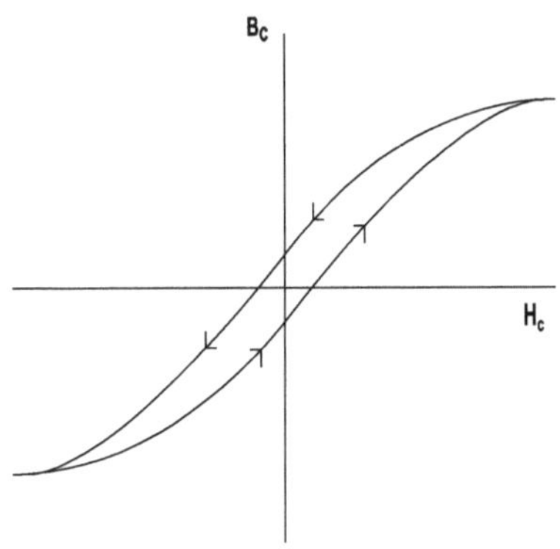

Figure 2.11. Magnetic hysteresis curve.

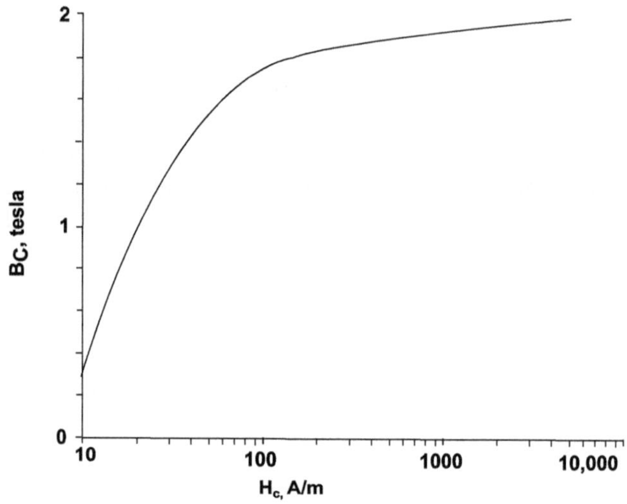

Figure 2.12. AC saturation curve.

density B_{max} that is at or perhaps slightly above the knee of the AC saturation curve. The required cross sectional area of the core can then be determined to be

$$A_c = \frac{\phi_{\text{design}}}{B_{\text{max}}} \tag{2.14}$$

The value of magnetic field at the design point, H_{max}, would be used to solve the equation

$$H_{\text{max}}\ell_c = Ni_{\text{max}} \tag{2.15}$$

The designer would need to select a wire size that is capable of carrying the design current. The designer would then select a core size with core length L_c that would allow N turns of this wire to pass through the center of the core. This is an iterative procedure, where a core is selected, the number of turns N is determined. The coil size is then determined, to see how well the coil would fit with the given core. If the coil size is too large to fit into the core window, a larger core is needed. If the coil size is smaller than the area available, a smaller core should be considered.

The flux linkage of the coil is the product of the coil turns and the flux linking the coil,

$$\lambda_{\text{coil}} = N\phi_{\text{coil}} \tag{2.16}$$

From these equations, at the design excitation level,

$$\lambda_{\text{design}} = N\phi_{\text{design}} = \frac{N^2 A_c B_{\text{design}}}{L_c H_{\text{design}}} i \tag{2.17}$$

Since inductance is defined as the ratio of flux linkage to current, the inductance of this device will be

$$L_{\text{design}} = \frac{N^2 A_c B_{\text{design}}}{L_c H_{\text{design}}} \tag{2.18}$$

The relationship between flux density B_{design} and magnetic field intensity H_{design} must be determined from the AC saturation curve. Note that this is specific to the design point, and the coil inductance will change as the current magnitude changes. A design that is strictly in the linear region of figure 2.12 will result in minimal change in inductance. A design that allows some level of saturation, perhaps in the range of H_c of around 100 A m^{-1} in this figure, might be able to provide a smaller, lighter inductor with acceptable performance for a given application.

Example 2.5. Determine the core flux in the magnetic core with the same dimensions as the one shown in figure 2.2. This core, however, is made with the magnetic material shown in the *B–H* curve of figure 2.10.

Solution. As the *B–H* curve is non-linear, the value of relative permeability is unknown and a graphical solution is necessary.

From example 2.1, the MMF applied to the magnetic circuit is $\mathscr{F} = Ni = 75$A-t. The effective core length is $L_C = 0.36$ m and the core cross sectional area is $A_c = 0.001\ \text{m}^2$.

The applied magnetic field is

$$\mathscr{H}_c = \frac{\mathscr{F}}{L_C} = \frac{75\ \text{amp} - \text{turns}}{0.36\ \text{m}} = 208.3\ \text{amp} - \text{turns/m}$$

The flux density is $\mathscr{B}_c = 1.44$ T. The core flux is then

$$\phi_c = A_c \mathscr{B}_c = 0.001\ \text{m}^2 \cdot 1.44\ \text{T} = 0.00144\ \text{webers}$$

Example 2.6. Repeat example 2.3, using the core material of figure 2.10 in place of the core material used in that example.

Solution. In this case, the MMF applied by the coil is distributed across the iron core and the air gap. An iterative solution is required to solve this problem. We start the iteration by assuming that there is no magnetic field within the iron core, so that all of the MMF appears across the air gap. Recall from example 2.3 that $\mathscr{F} = Ni = 75\ \text{A} - \text{t}$ and the air gap reluctance is $\mathfrak{R}_g = 1.912 \times 10^6$ amps/weber.

Iteration 1. The MMF across the air gap is $\mathscr{F} = \mathscr{F}_g = 75\text{A} - \text{t}$. The iteration 1 core flux is then

$$\phi_{c(1)} = \frac{\mathscr{F}_g}{\mathfrak{R}_g} = 39.23 \text{ microwebers}$$

From this, the iteration 1 core flux density is

$$\mathscr{B}_{c(1)} = \frac{\phi_c}{A_c} = 0.039226 \text{ T}$$

From figure 2.10, the core magnetic field is

$$\mathscr{H}_{c(1)} = 18 \text{ amp} - \text{turns/meter}$$

Iteration 2. The iteration 2 air gap MMF is

$$\mathscr{F}_{g(2)} = \mathscr{F} - \mathscr{H}_{c(1)}\ell_C = 75 - 18 \cdot 0.36 = 68.52$$

Using this new air gap MMF estimate in the same equations as in iteration 1, the core flux density in iteration 2 is

$$\mathscr{B}_{c(2)} = 0.03584 \text{ T}$$

The iteration 2 core magnetic field is then

$$\mathscr{H}_{c(2)} = 17.92 \text{amp} - \text{turns/meter}$$

Final solution. Following several further iterations, the solution converges to

$$\phi_c = 35.85 \text{ microwebers}$$

Core loss: Losses occur in the iron core from two different phenomena—hysteresis and eddy currents. Hysteresis loss is due to the energy that is required to re-orient the magnetic domains as the magnetic field changes. It is described by the area within the hysteresis curve shown in figure 2.11. Each cycle of AC excitation results in an energy loss in the magnetic core proportional to the area enclosed by the two paths of the curve. Magnetic materials intended for AC applications are selected and processed to minimize this area.

The second form of core loss is eddy current loss. Eddy currents in the core are induced by the changing flux field in the core. Because the core material is electrically conductive, current will flow in the core as well as in the coil. The I^2R losses in the core create heat and reduce the efficiency of the device.

Eddy current loss can be reduced by laminating the core. Thin laminations are arranged so the core flux travels through the core without crossing between laminations. An insulating material is present between laminations, and the laminations are arranged so that they disrupt the flow of electricity. Figure 2.13 shows the flux paths and current paths in solid cores and laminated cores. Laminations are effective in limiting eddy current loss in the lower frequency levels

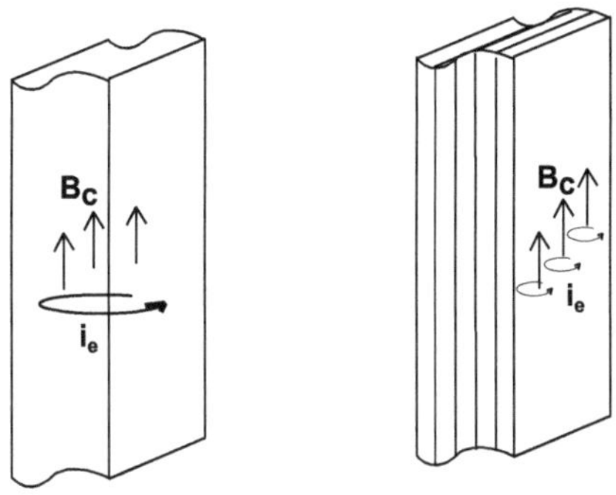

Figure 2.13. Eddy current paths in solid and laminated magnetic cores, showing the reduced eddy current path due to the laminations.

commonly used for electric power transmission (50 Hz, 60 Hz, 400 Hz). Advanced nanotechnology is being used to develop improved core materials for high frequency applications.

The insulating material between laminations has a finite thickness, and as a result, the effective cross sectional area of the core is reduced. In this text, we define the cross sectional area A_c to be the effective cross sectional area, and for laminated cores, it is reduced as appropriate to account for the impact due to the insulating layers.

Finite element analysis: The approximate magnetic circuit analysis discussed in this chapter provides a good starting point for electromagnetic device design. These designs are then refined through detailed magnetic field analysis using finite element analysis. A skilled designer using these advanced techniques can develop a precise and accurate design prior to the construction of a prototype device. In the case of high power machines, it is not practical to build a prototype at all, and the design must be highly accurate so that these multi-million dollar machines function as expected when they are commissioned.

Summary: Magnetic materials with high relative permeabilities are used in a variety of important electromagnetic devices. Two classes of these machines are the primary focus of this book—transformers and rotating machines.

Questions

1. Do a web search for manufacturers of grain oriented electric steel. Download one of their data sheets. Record the name of the manufacturer, the type of steel. Include a sketch of the *B–H* curve given for that product.
2. Some manufacturers provide data in units of oersteds for magnetic field intensity, and kilogauss for flux density. How do these units relate to the MKS units of amps/meter and tesla.
3. Manufacturers make a variety of pre-made cores for small transformers and inductors. Find a manufacturer of a specific type of cores, such as E cores or

toroidal cores, and describe the applications that the manufacturer intends for that product.

4. Powdered ferrite cores are used in some applications. Find a powdered core product, and describe the applications that the manufacturer intends for their powdered cores.

Problems

1. A ferromagnetic core is shown below. The relative permeability of the core is $\mu_r = 2400$.
   a. Find the effective core length for this core, and the cross sectional area of the core.
   b. What is the magnitude of the coil current is required to create a flux density of 1.25 T in the core?

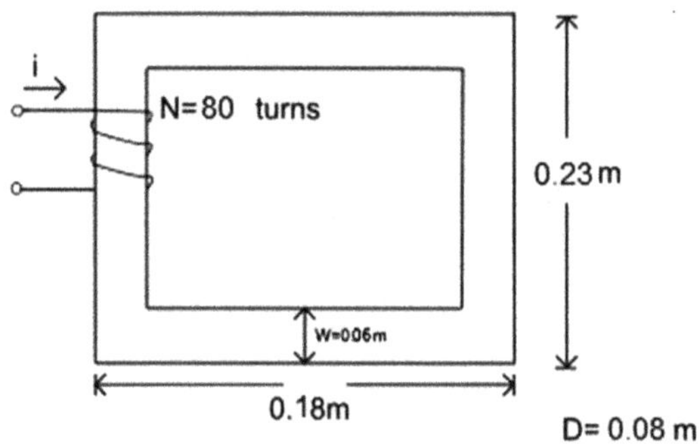

2. A magnetic core with an air gap is shown below. $\mu_r = 2100$. Find the core flux when the coil current is $i = 5$ A. The fringing factor for the gap is $\eta = 0.06$.

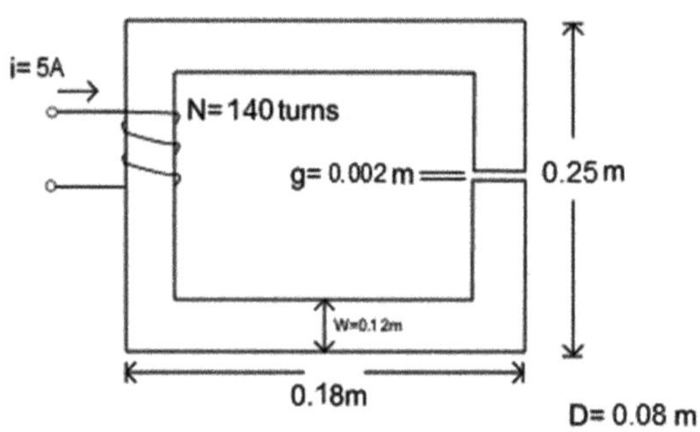

3. A magnetic circuit is shown below. The iron core has a relative permeability of $\mu_r = 1800$. The core depth $D = 0.20$ m. The current in the coil is $i = 2.4$ A. Find the core flux.

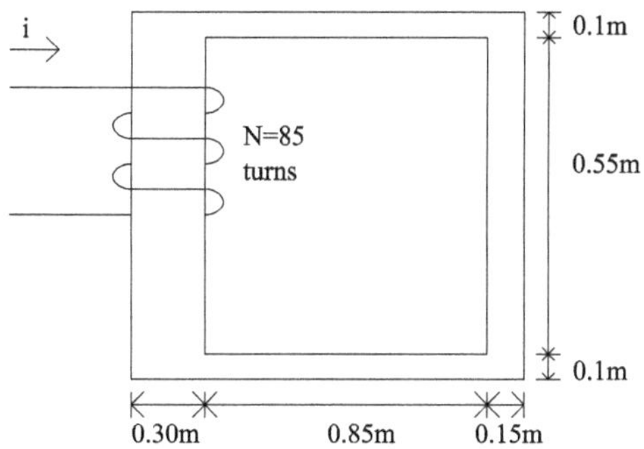

4. A common type of core is shown below, where the coil is wound on the center leg of a three leg core.

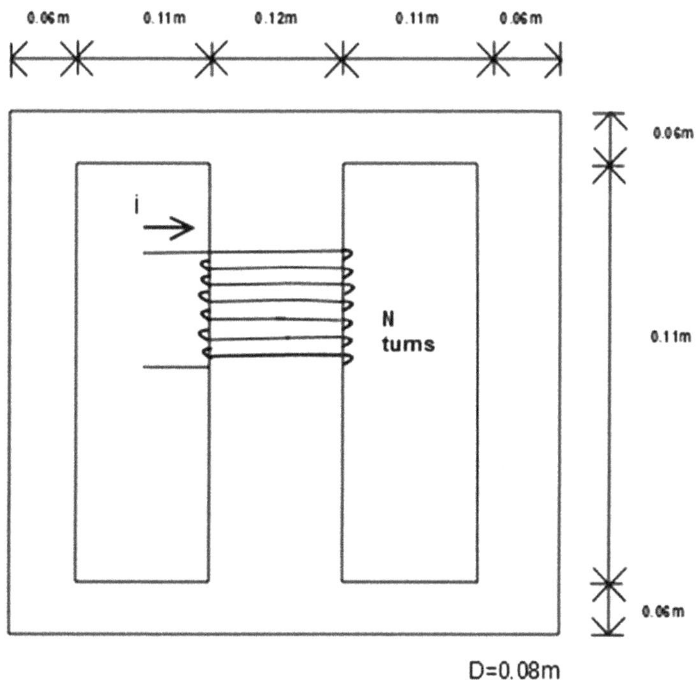

The relative permeability of the core is $\mu_r = 1800$.

a. Calculate the reluctance for each leg of the core.
b. A flux level of 12 milliwebers is desired in the center leg of this core for a current of 2.5 A. Determine the number of turns N required.
c. For this number of turns and current, calculate the flux linking the coil.
d. Determine the effective inductance of the coil for this operating point.

5. The *B–H* curve shown below is being used in a rectangular core whose dimensions are given in problem 1.

a. For an MMF of 20 amp-turns, determine the flux density from the *B–H* curve.
b. For this flux density, what is the core flux?
c. A family of inductors is being designed, all intended to work at an MMF of 20 amp-turns. Four inductors will be designed, with 20, 40, 80 and 160 turns, respectively. For each case, determine the coil current for each design that will give an MMF of 20 amp-turns.
d. For each case, determine the coil inductance at the design current for each of these coils.

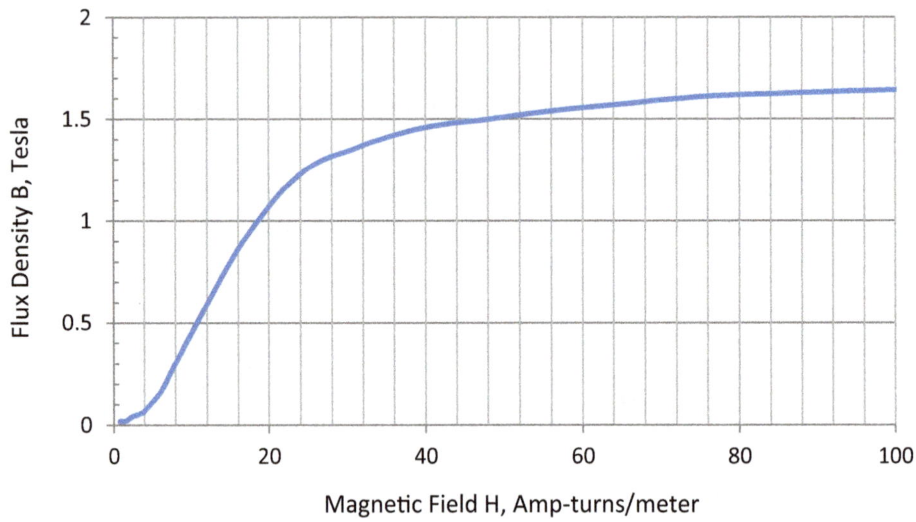

IOP Publishing

Electromechanical Machinery Theory and Performance

Thomas Ortmeyer

Chapter 3

Single phase transformers

Transformers play an important role in all of our lives. Transformer applications range from high power units that push the limits of transportation by rail, to small signal transformers that fit on a printed circuit board. While we often think of transformers as operating at the power frequency, there are many high frequency applications of transformers for both signals and power.

Transformers provide a convenient, energy efficient, cost effective means to change voltage and current levels in AC systems. They also can provide electrical isolation between the primary and secondary windings, to provide both human and equipment safety, as well as to allow one of the windings to be connected into a high voltage circuit while the other is ground referenced. They are used as an impedance matching device in some circuits, and to couple stages in amplifier circuits. Examples of several of these types of transformers are shown in figure 3.1.

3.1 Single phase two winding transformer

In its most basic form, a transformer consists of two coils wrapped around a single iron core, as shown in figure 3.2. The core is continuous, without an air gap. A two winding transformer is similar to the single winding iron core device discussed in chapter 2, with an additional winding included on the core. The two windings are often referred to as the primary and secondary windings. An alternative is to refer to them as the high and low side windings (often abbreviated to hi and lo). Formally, the term ‘primary’ refers to the winding the power enters, while ‘secondary’ refers to the winding that delivers power from its terminals. The alternate designations high and low side windings refer to the respective winding voltages.

In figure 3.2, the transformer windings are labeled as primary (‘p’) and secondary (‘s’). The primary current reference direction is into the coil (relative to voltage), and the secondary current reference direction is out of the coil. Using the right hand rule, you can see that positive primary current creates core flux in the clockwise direction, while positive secondary current creates core flux in the opposite direction.

doi:10.1088/978-0-7503-1662-0ch3

(a) Pole mount distribution transformer (b) Three phase air cooled transformer

Figure 3.1. Practical examples of transformers.

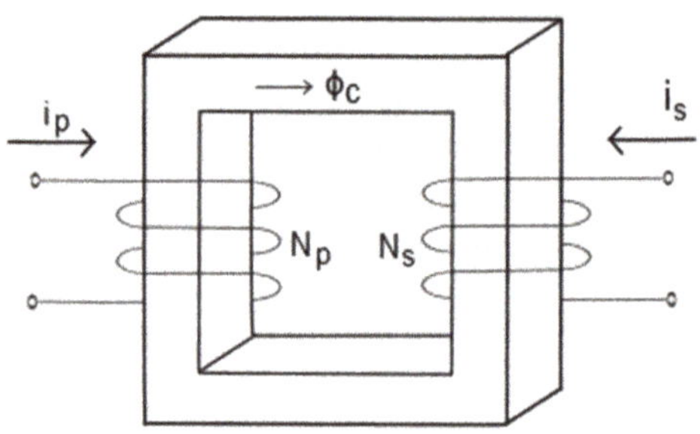

Figure 3.2. Two winding transformer conceptual diagram.

The lower case values i_p and i_s refer to time domain instantaneous quantities.

If the secondary is open circuited, $i_s = 0$. From chapter 2, the MMF in the core due to the primary current is

$$F_p = N_p i_p = H_c \ell_c = \frac{\phi_c \ell_c}{A_c \mu_r \mu_o} \tag{3.1}$$

In equation (3.1), N_p is the number of turns of coil p, H_c is the magnetic field intensity of the core, l_c is the effective core length, A_c is the effective core cross sectional area, μ_r is the relative permeability of the iron core material, and μ_o is the permeability of air.

Positive current flow in the secondary winding will counteract this MMF. When both primary and secondary current flow, the net core MMF is

$$F_{\text{core}} = F_{\text{p}} - F_{\text{s}} = N_{\text{p}} i_{\text{p}} - N_{\text{s}} i_{\text{s}} = \frac{\phi_{\text{c}} \ell_{\text{c}}}{A_{\text{c}} \mu_{\text{r}} \mu_o} \tag{3.2}$$

Here, N_s is the number of turns of the secondary coil. Note that setting $i_s = 0$ in equation (3.2) yields the special case of equation (3.1).

This core flux ϕ_c will link both coils. The core flux linking the primary coil will be

$$\lambda_{\text{p(core)}} = N_{\text{p}} \phi_{\text{c}} \tag{3.3}$$

and the core flux linking the secondary coil will be

$$\lambda_{\text{s(core)}} = N_{\text{s}} \phi_{\text{c}} \tag{3.4}$$

The induced voltage on a coil is the time derivative of the magnetic core flux linking that coil. We will designate this component of the total coil voltage as e_p and e_s for the primary and secondary coils respectively. The term '*e*' signifies electromotive force, commonly abbreviated emf. The term emf is usually reserved for a voltage that is induced onto a coil by magnetic fields.

For the transformer of figure 3.2, the primary and secondary emfs are

$$\begin{aligned} e_{\text{p}} &= \frac{d\lambda_{\text{p(core)}}}{dt} = N_{\text{p}} \frac{d\phi_{\text{c}}}{dt} \\ e_{\text{s}} &= \frac{d\lambda_{\text{s(core)}}}{dt} = N_{\text{s}} \frac{d\phi_{\text{c}}}{dt} \end{aligned} \tag{3.5}$$

Equation (3.5) shows that the relationship between these two emfs is:

$$e_{\text{p}} = \frac{N_{\text{p}}}{N_{\text{s}}} e_{\text{s}} \tag{3.6}$$

The ratio $\frac{N_p}{N_s}$ is referred to as the turns ratio of the transformer. It is often given the symbol '*a*':

$$a = \frac{N_{\text{p}}}{N_{\text{s}}} \tag{3.7}$$

So that

$$e_{\text{p}} = a e_{\text{s}} \tag{3.8}$$

Example 3.1. An iron core has an effective cross sectional area of $A_c = 0.01$ square meters. A coil of $N_p = 1800$ turns is connected to a 60 Hz voltage source $e_p = \sqrt{2}(7200)\cos(2\pi 60t) = \sqrt{2}(7200)\cos(377t)$ V.

(a) Determine the core flux density.

From equation (3.5),

$$e_p = \frac{d\lambda_{p(core)}}{dt} = N_p \frac{d\phi_c}{dt}$$

So

$$\frac{d\phi_c}{dt} = \frac{e_p}{N_p} = \sqrt{2}\,(4.0)\cos(377t)$$

Integrating the cosine function,

$$\phi_c = \sqrt{2}\frac{4.0}{377}\sin(377t) + C$$

The constant C from the integration would equal zero in the steady state. Converting the sine function to a cosine function,

$$\phi_c = 0.0150\cos\left(377t - \frac{\pi}{2}\right)$$

The flux density would then be

$$B_c = \frac{\phi_c}{A_c} = 1.50\cos\left(377t - \frac{\pi}{2}\right)\text{ T}$$

The peak flux density would be 1.50 T, and the flux and flux density waves would lag the voltage wave by $\frac{\pi}{2}$ radians or 90°. As 60 Hz is a common frequency, it is worth remembering that the frequency of 60 Hz gives an angular frequency of $\omega = 377$ radians per second.

(b) Determine the number of secondary turns required for a secondary voltage of 120 V RMS when the primary voltage is 7200 V RMS.

One way to find the secondary turns N_s is to use equation (3.6), $e_p = \frac{N_p}{N_s}e_s$. Formally,

$$\sqrt{2}\,(7200)\cos(2\pi 60t) = \frac{1800}{N_s}\sqrt{2}\,(120)\cos(2\pi 60t)$$

Or

$N_s = 30$ turns. Note that the equation shows that these two voltages must be in phase.

(c) Determine the turns ratio of the transformer.

From equation (3.7), the turns ratio a is

$$a = \frac{N_p}{N_s} = \frac{1800}{30} = 60$$

As a is a ratio, it is unitless.

3.2 The ideal transformer

It is useful to define the concept of an 'ideal' transformer. As implied by the term, ideal transformers do not actually exist. The ideal transformer, however, is convenient to use for modeling and analysing the real transformer. It is a part of the real transformer model that will be developed, and there are cases where an analysis conducted with an ideal transformer model is sufficient for a given study (and there are just as many or more cases where it is not).

In an ideal transformer, the level of MMF required to generate the core flux is vanishingly small. Equation (3.2) then becomes

$$0 = N_p i_p - N_s i_s \tag{3.9}$$

or

$$i_p = \frac{N_s}{N_p} i_s = \frac{i_s}{a} \tag{3.10}$$

A quick analysis shows that the instantaneous power going into the primary ($e_p i_p$) equals the instantaneous power coming out of the secondary ($e_s i_s$). The ideal transformer neither generates, consumes, nor stores energy.

Figure 3.3 introduces the schematic diagram of the ideal transformer. It shows the turns ratio as either a:1 or N_p:N_s. It also shows the polarity of the transformer, through the dots that are at the top of both coils in the figure. The polarity defines the relationship between primary and secondary quantities:

- If the secondary winding voltage e_s is positive when measured from the polarity mark to the non-polarity side of the coil, e_p will be positive when measured across the coil from polarity to non-polarity.
- If i_s is positive coming out of the polarity mark, i_p will be positive going into the polarity mark.

The vertical lines between the coils represent the iron core. Note that these vertical lines are not always drawn in transformer diagrams.

The preceding equations and figure 3.3(a) involve instantaneous time domain quantities. Transformers, however, work with AC quantities. They cannot tolerate

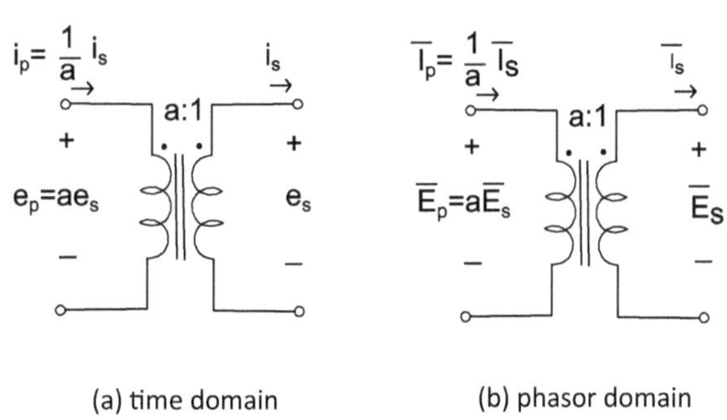

Figure 3.3 Ideal transformer.

direct currents (DC) for very much time without saturating the iron core. The ideal transformer model is not valid when the core is saturated.

Transformers are often analysed in the phasor domain. The phasor equivalent of figure 3.3(a) is shown in figure 3.3(b). In phasor terms,

$$\overline{E}_p = a\overline{E}_s$$
$$\overline{I}_p = \frac{1}{a}\overline{I}_s \tag{3.11}$$

In this book, phasors are noted by overbars. The *magnitude* of a phasor does not have the overbar. The angle of the phasor can be given in units of degrees or radians. The magnitude of the phase is the root mean square (RMS) value of the sinusoidal function in the time domain.

Therefore, $\overline{E}_p = E_p/\underline{\delta_p}$ denotes a phasor of magnitude E_p and angle δ_p.

One characteristic of an ideal transformer is that it changes the apparent impedance of a load. Apparent impedance is defined as the ratio of voltage across a device or circuit to current entering the same device or circuit.

Figure 3.4 shows an ideal transformer fed by a source with a load impedance $\overline{Z}_L$ connected on the secondary. As viewed from the secondary terminals, the apparent impedance of the load is the actual impedance

$$\overline{Z}_{S(apparent)} = \frac{\overline{V}_S}{\overline{I}_S} = \overline{Z}_L \tag{3.12}$$

So the apparent impedance of a load of constant impedance is the impedance itself. On the primary side, the apparent impedance is the ratio of the primary voltage to the primary current,

$$\overline{Z}_{p(apparent)} = \frac{\overline{V}_p}{\overline{I}_p} = \frac{a\overline{V}_s}{\overline{I}_s/a} \tag{3.13}$$

or

$$\overline{Z}_{p(apparent)} = a^2\overline{Z}_{s(apparent)} \tag{3.14}$$

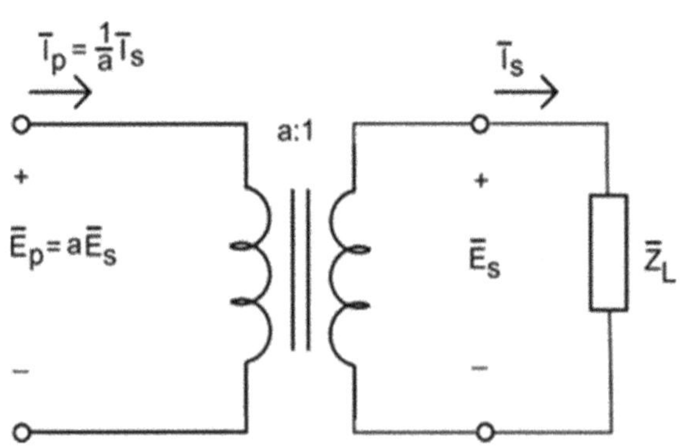

Figure 3.4. Ideal transformer with constant load impedance $\overline{Z}_L$.

This impedance shifting capability of transformers is used in audio and radio frequency applications. Impedance matching transformers are also used at frequencies up to approximately 1 MHz in industrial plasma systems, with applications including metal extraction, lighting, coating, etching, and cutting.

Example 3.2. An ideal transformer is shown below. The voltage applied to the load is 120 V RMS.

a. Using phasor quantities, determine the voltage and current at the primary terminals of the transformer, using the ideal transformer theory.

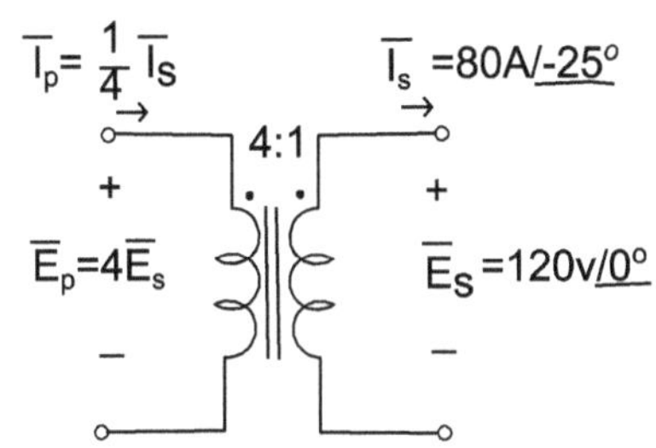

$$\bar{E}_p = 4\bar{E}_s = 480v \,\underline{/0^\circ}$$

$$\bar{I}_p = \frac{1}{4}\bar{E}_s = 20A \,\underline{/-25^\circ}$$

$$\bar{Z}_{p(apparent)} = \frac{480\text{ V}}{20\text{ A}} = 24\ \Omega$$

or

$$\bar{Z}_{p(apparent)} = a^2 1.5 = 24\ \Omega$$

b. Calculate the secondary real and reactive power and primary real and reactive power for this case.

$$P_s = V_s I_s \cos(\delta_s - \phi_s) = 120 \cdot 80 \cos(+25^\circ) = 8700\text{ W}$$
$$Q_s = V_s I_s \sin(\delta_s - \phi_s) = 120 \cdot 80 \sin(+25^\circ) = 4057\text{ VARs}$$
$$P_p = V_p I_p \cos(\delta_p - \phi_p) = 480 \cdot 20 \cos(+25^\circ) = 8700\text{ W}$$
$$Q_p = V_p I_p \sin(\delta_p - \phi_p) = 480 \cdot 20 \sin(+25^\circ) = 4057\text{ VARs}$$

Since there are no real or reactive power elements in the ideal transformer, the input and output real and reactive powers are equal.

Example 3.3. You have loudspeaker that has an impedance of 8 ohms at audio frequencies. You want to drive this by a power amplifier that is rated 50 V and 0.5 A.

a. If you connect the amplifier directly to the speaker, will it reach rated voltage or rated current first?

At rated current,

$$V_{\text{amp}} = V_{\text{speaker}} = 8\ \Omega\ (0.5\ \text{A}) = 4\ \text{V}$$

4 V is within the rating of the amplifier, so this setting is acceptable.

At rated voltage,

$$I_{\text{amp}} = I_{\text{speaker}} = \frac{50\ \text{V}}{8\ \Omega} = 6.25\ \text{A}$$

6.25 A greatly exceeds the speaker's current rating, so the amplifier will either shut down or fail if it attempts to drive this speaker at 40 V.

b. An impedance matching transformer will be connected between the amplifier and the speaker. What is the ideal transformer ratio of this transformer?

It is desired that the power amplifier see an apparent impedance of

$$Z_{\text{p(apparent)}} = \frac{50\ \text{V}}{0.5\ \text{A}} = 100\ \Omega.$$

Therefore,

$$100\ \Omega = a^2 \cdot 8\ \Omega$$

So the transformer ratio should be $a = \sqrt{\frac{100}{8}} = 3.54$

c. What is the maximum power that can be delivered to the speaker for these two connections?

Without the transformer,

$$P_{\text{speaker}} = 4\ \text{V} \cdot 0.5\ \text{A} = 2\ \text{W}$$

With the transformer,

$$P_{\text{amp\_out}} = 50\ \text{V} \cdot 0.5\ \text{A} = 25\ \text{W}$$

and at the speaker,

$$P_{\text{speaker}} = 14.14\ \text{V} \cdot 1.77\ \text{A} = 25\ \text{W}$$

This transformer will therefore allow for a much better utilization of the power amplifier. The speaker would also need to be checked to verify that it is rated for at least 25 W.

3.3 The real transformer

As could be expected, actual transformers differ from ideal transformers in several ways. That said, a well designed transformer operating within its intended operating range comes close to ideal performance.

The real transformer model accounts for core magnetization, core loss, winding resistance, and leakage flux. The first two effects are properties of the core itself, while the second two are properties of individual windings. These two aspects will be treated separately in the next two subsections.

3.3.1 Core magnetization and core loss

The ideal transformer assumes infinite permeability which results in a negligible amount of current required to magnetize the core. While relative permeability of iron and other ferromagnetic materials is large, it is not infinite. As a result, a certain amount of current is required to create the magnetic flux in the transformer core. This current is commonly referred to as the magnetizing current.

The second issue with core magnetization is that the relative permeability of the core can be considered to be constant only up to a certain level of magnetic flux density, where the ferromagnetic core will start to saturate. The following analysis assumes that the transformer is being operated below its saturation point.

Linear magnetization

When the transformer is being operated so that relative permeability is essentially constant, equation (3.2) holds. Consider the situation when the transformer secondary is open circuited. Then

$$N_p i_p = \frac{\phi_c \ell_c}{A_c \mu_r \mu_o} \tag{3.15}$$

With the secondary open circuited, the primary current goes entirely to magnetize the core. The flux linking the primary coil is

$$\lambda_p = N_p \phi_c = N_p^2 \frac{A_c \mu_r \mu_o}{\ell_c} i_p \tag{3.16}$$

Magnetizing inductance

In circuit terms, it is inductance that relates current to flux linkage. Therefore, the magnetizing inductance can be defined as

$$L_{mp} = \frac{N_p^2 A_c \mu_r \mu_o}{\ell_c} \tag{3.17}$$

The subscript 'p' is included as this is the magnetizing inductance as viewed from the primary coil. If the magnetizing inductance is to be considered as being on the secondary side (the choice is strictly a matter of convenience), the value of this inductance would be

$$L_{ms} = \frac{N_s^2 A_c \mu_r \mu_o}{\ell_c} = \frac{L_{mp}}{a^2} \tag{3.18}$$

The induced voltage across this coil will be the time derivative of the flux linkage

$$e_p = \frac{d\lambda_p}{dt} \tag{3.19}$$

In steady state sinusoidal conditions, this equation can be written in phasor terms as

$$\overline{E}_p = j\omega L_{mp}\overline{I}_{mp} \tag{3.20}$$

In this linear case with no iron core saturation, the model predicts that the transformer will draw a sinusoidal magnetizing current in the steady state when a sinusoidal voltage is applied. As discussed in the next section, the core is actually non-linear, and the magnetizing current will not be sinusoidal.

Example 3.4. For the iron core of example 3.1, the effective core length is 0.5 m, and the relative permeability of the core is 1400.

(a) Find the RMS value of the MMF required to excite the core at a flux density level of 1.5 T.

From equation (3.1) and example 3.1, the magnetic field intensity is

$$F_p = H_c\ell_c = \frac{B_c\ell_c}{\mu_r\mu_o} = \frac{1.50\cos\left(377t - \frac{\pi}{2}\right)(0.5)}{(1400)(4\pi \times 10^{-7})} = 426.3\cos\left(377t - \frac{\pi}{2}\right) \text{ amp-turns}$$

(b) From equation (3.2), find the level of RMS primary current required to excite this core at a flux density of 1.5 T. (Hint: this calculation will work if you assume that the secondary current $i_s = 0$ in equation (3.2), and all of the primary current is going to excite the core.)

From equation (3.2), $F_{core} = F_p - F_s = N_p i_p - N_s i_s$. With $i_s = 0$, the instantaneous primary current is

$$i_p = \frac{F_p}{N_p} = 0.237\cos\left(377t - \frac{\pi}{2}\right) \text{ A}$$

The RMS current is then

$$I_p = \frac{0.237}{\sqrt{2}} = 0.167 \text{ A}$$

(c) From a circuits point of view, the MMF required to excite the core is created from a current that appears to flow through an inductor connected in parallel with the coil. The inductance of this apparent inductor is the ratio of flux linkage to current. What is the inductance of this iron core, as viewed from the primary side of the transformer?

From example 3.1, the core flux is

$$\phi_c = 0.0150 \cos\left(377t - \frac{\pi}{2}\right)$$

The flux linking the primary winding is then

$$\lambda_p = N_p\phi_c = 27 \cos\left(377t - \frac{\pi}{2}\right)$$

The apparent inductance is then

$$L_{core} = \frac{\lambda_p}{i_p} = 113.9 \text{ henries}$$

(d) If the frequency of the applied to the transformer is 60 Hz, what is the inductive reactance of this inductance, again as viewed from the primary winding?

$$X_{core} = \omega L_{core} = (2\pi 60)(113.9) = 42\,950\ \Omega \ \textit{or}\ 42.95\ \text{k}\Omega$$

Non-linear core operation

In actuality, it is not economical to design power transformers or many of the other transformers to operate strictly in the linear magnetic region. In modern transformer designs, the magnetic core flux will go into some level of saturation at peak flux linkages. This is illustrated in excitation curve shown in figure 3.5. Due to the iron core saturation and hysteresis, the excitation current now will be non-sinusoidal, even when the exciting voltage is sinusoidal. The full transformer equivalent circuit ignores this non-linearity, which allows the transformer to be analysed as a linear circuit. This provides a significant simplification of the analysis. This must be remembered when applying this transformer model—there are many situations where this approximation gives an appropriate prediction of transformer performance. There are, however, some instances where the full equivalent circuit is not sufficiently accurate. An example is when knowledge of the transformer harmonic currents is of interest. Another situation is during transformer energization—when the transformer is newly connected to the system after being off-line. In these cases, more accurate non-linear models are needed.

A transformer can be tested by applying a sinusoidal voltage to one winding, and measuring the resulting rms current flow. A plot of these results is referred to as the AC saturation curve for the transformer. A representative AC saturation curve is shown in figure 3.5. Also shown in figure 3.5 is the straight line approximation of the magnetizing reactance of the excitation curve. The two curves intersect at rated winding voltage, and the slope of the curve is the magnetizing reactance of the transformer. The difference between these two curves is a measure of the error due to

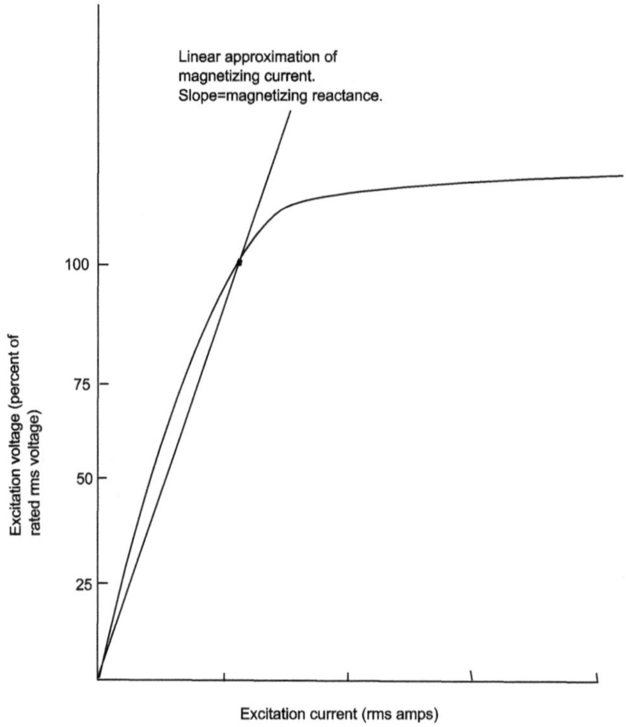

Figure 3.5. AC excitation curve for an unloaded transformer.

this assumption. As transformers typically operate between 95% and 105% of rated voltage, the error in this example case remains small.

Core loss and core loss resistance

As discussed in chapter 2, the iron core is subject to both hysteresis loss and eddy current loss. Both of these loss components are non-linear, and are both voltage and frequency dependent. The core loss tends to be proportional to frequency while the eddy current loss tends to be proportional to the square of the frequency. In transformers, these losses are also proportional to the square of the excitation voltage, and the core loss is represented by a fixed resistance in parallel with the magnetizing reactance of the transformer.

When a transformer is operated over a small range of AC voltage and frequency, core loss can be represented by a fixed resistance that is connected in parallel with the magnetizing inductance. This provides a reasonable model for AC power transformers, where the normal range of voltage magnitude variation is ±5% and the normal frequency variation is less than ±1%.

The transformer equivalent circuit considering with the magnetizing reactance X_{mp} and core loss resistance R_{cp} added to the ideal transformer is shown in figure 3.6.

3.3.2 Winding resistance and leakage reactance

Individual windings on the transformers also have non-ideal effects of coil resistance and leakage inductance.

Leakage inductance: While the large majority of flux in a transformer is confined to the core, each coil has a small component of flux that links only that coil and takes a path that includes the core, the surrounding insulating material, and perhaps the outside casing and/or the bracing structures holding the core in place. A conceptual diagram of the leakage flux path is shown in figure 3.7. This leakage flux can be modeled by an inductance that is connected in series with the winding in the equivalent circuit.

Coil resistance: There are several factors that make the coil resistance different than the DC resistance that can be measured at the coil terminals. These are

- skin effect;
- winding temperature;
- stray losses.

The skin effect within the coil is due to the internal inductance of the coil. This inductance forces alternating currents away from the center of the conductor, resulting in a non-uniform current distribution and a higher effective resistance to these alternating currents. This impact increases with frequency—it is small but not

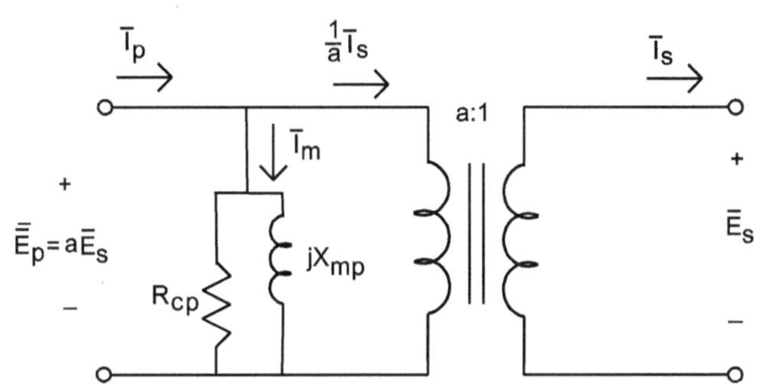

Figure 3.6. Transformer model with magnetizing branch included.

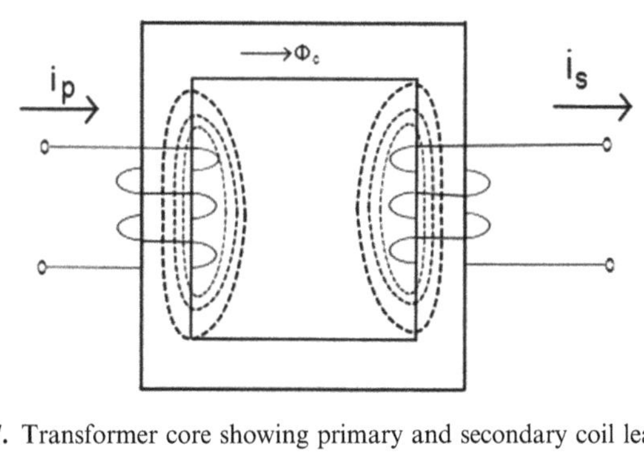

Figure 3.7. Transformer core showing primary and secondary coil leakage flux.

negligible at power system frequencies, and can become a dominating factor at high frequencies.

The coil resistance also changes with winding temperature. This change can be predicted by well known formulas. When manufacturers specify transformer coil resistance, it will be given for the rated operating temperature. It is common practice to use this specified value for system studies.

Stray losses refer to losses outside the coil that are due to the leakage flux. These vary with the load current rather than the excitation voltage. From the terminals, they appear to be in series with the coil resistance, and are therefore lumped into a single equivalent resistance for each coil in the transformer model.

3.3.3 The full transformer equivalent circuit

The ideal transformer model must be modified by these core and coil effects to obtain a more accurate transformer equivalent circuit. This equivalent circuit is shown in figure 3.8. This complete equivalent circuit includes the ideal transformer plus three resistance and three reactance elements. The resistances R_p and R_s represent the equivalent coil resistance of the primary and secondary windings, and the resistance R_c represents the core loss of the transformer, as viewed from the primary winding. The reactance X_{mp} represents the transformer magnetizing inductance as referred to the primary coil, and the reactances X_p and X_s represent the leakage inductance of the primary and secondary coils. The phasor equivalent circuit uses the equivalent impedances of the inductances. Examination of the equivalent circuit shows that R_p and jX_p carry the primary winding current, while R_s and jX_s carry the secondary winding current. The magnetizing voltage appears across the R_{cp} and jX_{mp} elements—neither carry the total primary or secondary current.

It is often convenient to represent all of the impedance elements of the transformer equivalent circuit on the same side of the ideal transformer. These could be on either the primary winding or the secondary winding. The impedance elements that are moved are referred to the other side by the turns ratio squared, as discussed in section 3.2. The resulting primary equivalent circuit is shown in figure 3.9, and the secondary equivalent circuit is shown in figure 3.10. In this figure,

$$R_{cs} = \frac{R_{cp}}{a^2} \quad X_{cs} = \frac{X_{cp}}{a^2} \tag{3.21}$$

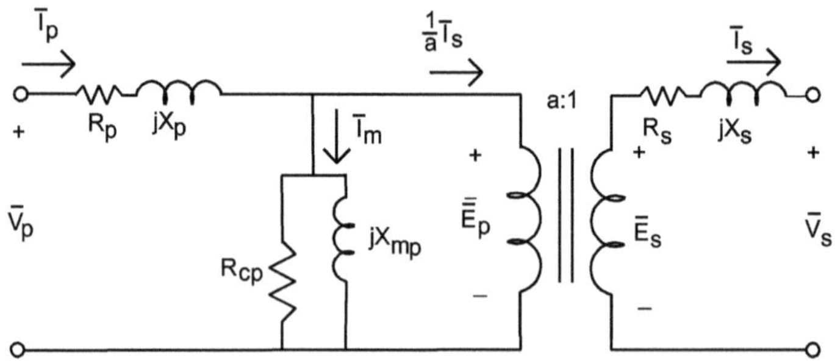

Figure 3.8. Full equivalent circuit of the transformer.

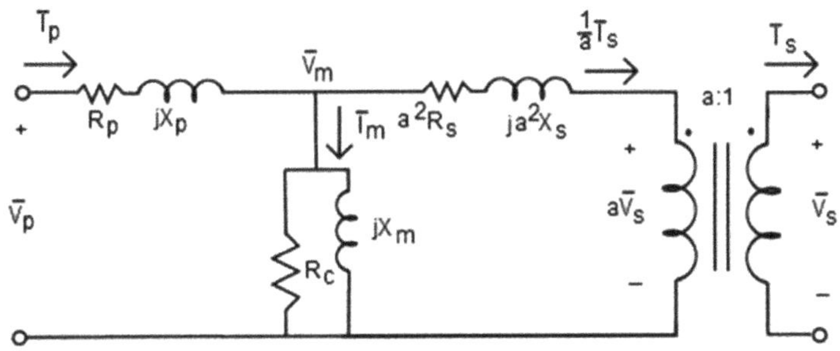

Figure 3.9. Full transformer equivalent circuit with impedances referred to the primary. Note that the magnetizing voltage V_m is defined for convenience.

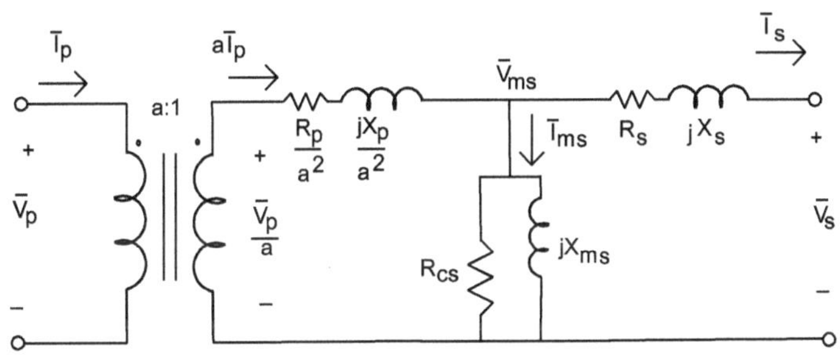

Figure 3.10. Full transformer equivalent circuit with impedances referred to the secondary.

These are both full equivalent circuits, and will give identical terminal results at the transformer terminals.

3.3.4 Simplified equivalent circuits

Cases often arise where a simplified approximate equivalent circuit can be used and still achieve acceptable accuracy. Figure 3.11 shows several of these simplified equivalent circuits. A common simplification is to combine both winding resistance and leakage inductance elements into single elements. This is shown in figure 3.11(a), where the lumped impedances are on the primary side of the magnetizing branch. The impedances are

$$R_{eq} = R_p + a^2 R_s \quad X_{eq} = X_p + a^2 X_s \tag{3.22}$$

This impedance can just as well be placed on the other side of the magnetizing branch.

A second simplification is to neglect the core loss resistance R_{cp}, as shown in figure 3.11(b). A third one, shown in figure 3.11(c), is to neglect the magnetizing reactance X_{mp}. The winding resistances and leakage reactances can also be neglected—with the result that the model now consists strictly of the ideal transformer.

Equivalent models can be developed with impedance represented on the secondary side.

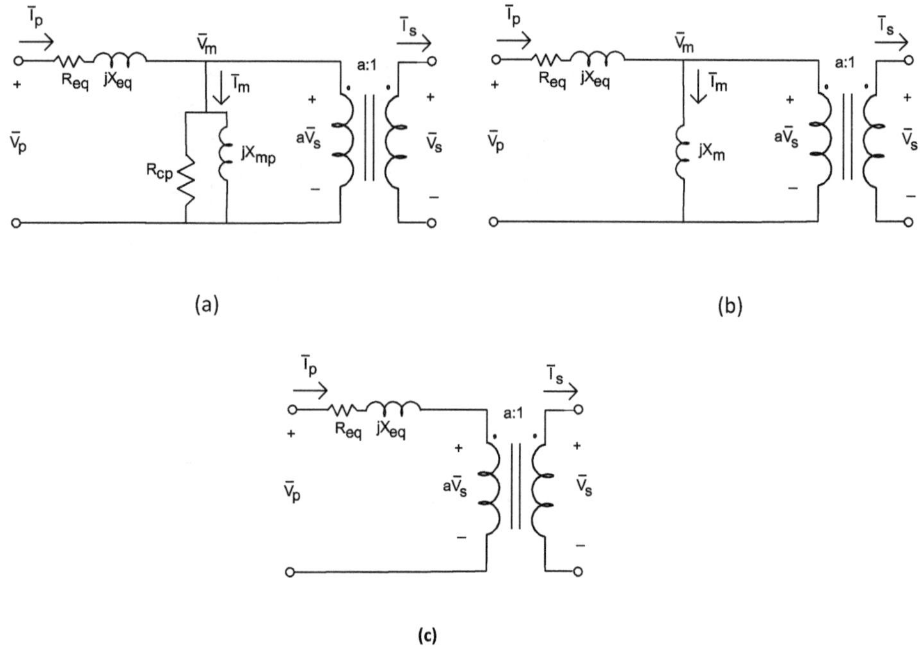

Figure 3.11. Three simplified transformer equivalent circuits. Primary side versions.

Each of these models is used by engineers to design and analyse systems that include transformers. One definition of a good engineer is one who knows which model to use for a given study. This knowledge generally comes from experience, and demonstrating to oneself that the use of a simplified model will provide sufficient accuracy in the particular case under study.

The full magnetizing branch must be included when transformer losses and efficiency are of interest, so either a full equivalent circuit or the circuit of figure 3.11(a) should be used. For system studies of voltage drop and short circuit analysis, the magnetizing branch is often neglected. On the other hand, significantly more detailed models than any of these equivalent circuits are used by transformer designers in developing the transformer designs.

Example 3.5. A transformer has the following characteristics:

$R_p = 1.20\ \Omega$	$X_p = 4.40\ \Omega$	$R_s = 0.012\ \Omega$	$X_s = 0.044\ \Omega$
$a = 10$	$R_{cp} = 12\ 800\ \Omega$	$X_{mp} = 3200\ \Omega$	

a. Draw the full transformer equivalent circuit with all elements referred to the primary winding.

The given values of R_p, X_p, R_c and X_m are as referred to the primary. The secondary winding resistance and leakage reactance need to be referred to the primary for this equivalent circuit.

$$a^2R_s = 10^2(0.012) = 1.2\ \Omega$$
$$a^2X_s = 10^2(0.044) = 4.4\ \Omega$$

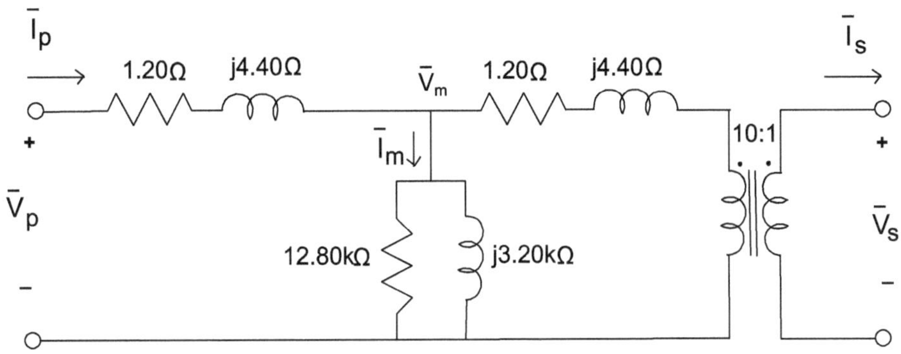

Full equivalent circuit, all elements referred to the primary.

b. Draw the full transformer equivalent circuit with all elements referred to the secondary winding.

The primary winding resistance and leakage reactance need to be referred to the secondary. The core loss resistance and magnetizing reactance also need to be referred to the secondary.

$$R_p/a^2 = (1.2)/100 = 0.012\ \Omega$$
$$X_p/a^2 = (4.4)/100 = 0.044\ \Omega$$
$$R_c/a^2 = (12800)/100 = 128.0\ \Omega$$
$$X_m/a^2 = (3200)/100 = 32.0\ \Omega$$

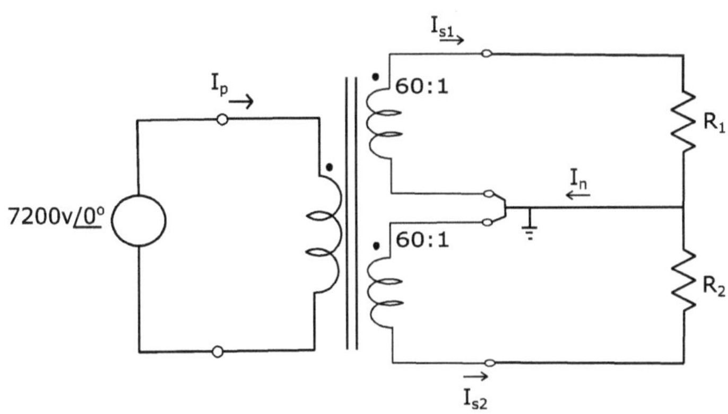

Full equivalent circuit, all elements refer to the secondary.

c. If a load is connected to the transformer, so that $\overline{I}_s = 60\text{A}\underline{/-20^\circ}$ when $\overline{V}_s = 120\text{V}\,\underline{/0^\circ}$, find the primary voltage and current that are supplying the transformer. Use the full equivalent circuit with all elements referred to the primary for this calculation.

In this equivalent circuit, the secondary current and voltage are first referred to the primary:

$$a\overline{V}_s = 10(120\ \text{V}\ \underline{/0^\circ}) = 1200\ \text{V}\ \underline{/0^\circ}$$
$$\overline{I}_s/a = (60\ \text{A}\ \underline{/-20^\circ})/10 = 6.0\ \text{A}\ \underline{/-20^\circ}$$

The magnetizing voltage $\overline{V}_m$ is defined in the equivalent circuit. From this circuit,

$$\overline{V}_m = a\overline{V}_s + (a^2R_s + ja^2X_s)\frac{\overline{I}_s}{a}$$

or

$$\begin{aligned}\overline{V}_m &= 1200\ \text{V}\ \underline{/0^\circ} + (1.20\ \Omega + j4.40\ \Omega)6.0\ \text{A}\ \underline{/-20^\circ} \\ &= 1215.8 + j22.3\ \text{V} = 1215.8\ \text{V}\ \underline{/1.1^\circ}\end{aligned}$$

There are several different ways to calculate the current flowing into the magnetizing branch. One method is:

$$\overline{I}_R = \frac{\overline{V}_m}{R_c} = \frac{1215.4\ \text{V}\ \underline{/1.1^\circ}}{12\ 800\ \Omega} = 0.0950 + j0.0017\ \text{A}$$
$$\overline{I}_x = \frac{\overline{V}_m}{jX_m} = \frac{1215.4\ \text{V}\ \underline{/1.1^\circ}}{j3200\ \Omega} = 0.0070 - j0.3799\ \text{A}$$

These are the currents in R_c and X_m, respectively. The total magnetizing current is then

$$\overline{I}_m = \overline{I}_R + \overline{I}_x = 0.1020 - j0.3782\ \text{A} = 0.392\ \underline{/-74.9^\circ}$$

The primary current is then

$$\overline{I}_p = \overline{I}_m + \frac{\overline{I}_s}{a} = 6.233\ \text{A}\ \underline{/-22.9^\circ}$$

The primary voltage is

$$\overline{V}_p = \overline{V}_m + (R_p + jX_p)\overline{I}_p$$

or

$$\overline{V}_p = 1215.8\ \text{V}\ \underline{/1.1^\circ} + (1.20 + j4.40)6.233\ \text{A}\ \underline{/-22.9^\circ} = 1234.2\ \text{V}\ \underline{/2.1^\circ}$$

The solution of the full equivalent circuit with values referred to the secondary will give the same results for the primary voltage and current.

3.4 Transformer ratings

Transformers are rated so that the transformer users can apply them effectively, so that they can be operated within their capability while avoiding unnecessary expense from overbuilding the unit. In order to facilitate the rating systems, specifications have been developed that define what the device ratings mean, and how they are determined. The most commonly used power transformer ratings are specified by the Institute of Electrical and Electronics Engineers (IEEE) or by the International Electrotechnical Commission (IEC). Power transformers ratings cover a wide range of issues, including but not limited to:

- steady state electrical capability;
- electrical transients;
- short circuit performance;
- construction;
- altitude;
- current/voltage waveshape;
- thermal rating.

Separate standards are available for liquid filled, dry type, overhead, pad-mounted, and underground transformers.

In this section, we will concentrate on the normal steady state ratings of a transformer. The typical steady state transformer ratings are voltage, volt-amps, frequency, temperature, and turns ratio.

Voltage rating: A voltage rating is supplied for each winding of the transformer. The ratios of these ratings give the turns ratio of the transformer—the turns ratio is not specified separately. The voltage ratings are given in terms of RMS values. Transformer voltages are generally stated in terms of nominal values. Transformers are designed to operate within a range around these nominal values—typically within ±5% of the nominal voltage rating. The voltage rating is determined by two primary factors—the capability of the winding insulation, and the need to avoid undue iron core saturation. The voltage rating applies to steady state conditions.

Volt-amp rating: The volt-amp rating of a single phase two winding transformer is the continuous volt-amps that the transformer can supply under standard conditions at rated secondary voltage without exceeding the thermal limitations of the transformer.

Frequency rating: The frequency rating is the nominal frequency at which the transformer is designed to operate. Again, the standards allow the transformer to operate within a range of frequencies near the rated frequency. Operation outside of this frequency range can be possible, but will often require careful analysis and derating of the transformer to avoid damage.

Temperature: The transformer core, coils, and insulation all have temperature limits that should not be exceeded. At certain temperatures, both core and coil metals will be damaged, through annealing or even melting. Insulating materials experience rapid aging at elevated temperatures.

Many power transformers are immersed in liquid. The liquid acts as both a cooling agent and an insulating agent. Elevated temperatures degrade the insulating properties of this liquid, in a similar manner as happens with solid insulation.

Another class of power transformers are dry type transformers. These are generally used inside buildings, where the transformer is protected from the weather, and flammable insulating oils would be not allowed.

Transformer designers think in terms of the increase in temperature of the coil and core caused by losses at the transformer rated point. The limiting factor in many designs is the temperature of the solid insulation on the conductors. In these cases, the transformer insulation capability less the temperature rise at the transformer rated point will give the maximum ambient temperature at which the transformer can operate. If operating at ambient temperatures above this value is required, the transformer must be derated to avoid overheating. Details should be worked out between the transformer owner and the transformer manufacturer, particularly when warranties are in place. With large power transformers, there is generally an expectation that the unit will last 40 years or more when it is operated within its design range.

Example 3.6. A 60 Hz transformer is rated at 4800 V:240 V, 800 kVA.

(a) Determine the turns ratio of the transformer

$$a = \frac{4800}{240} = 20$$

(b) Determine the rated primary and secondary currents for this transformer.

$$I_{\text{rated primary}} = \frac{800 \text{ kVA}}{4800 \text{ V}} = 166.7 \text{ A}$$

$$I_{\text{rated secondary}} = a \cdot I_{\text{rated primary}} = 3333\text{A}$$

In calculating the primary amps, note that the ‘k’ in kVA is interpreted as 10^3 in finding the resulting current in amps. This notational method for ‘kilo’ is followed in this book, as is ‘mega’ meaning 10^6.

Most electric power equipment will have a nameplate mounted on the device. A typical nameplate is shown in figure 3.12. This nameplate shows the transformers manufacturer and model number, the electrical and thermal ratings of the device, the transformer series impedance expressed in per cent, weight, and the transformer connection diagram. Additional information is included in a product specification sheet, installation manual and/or operation manual. The nameplates and other documents are important in specifying, installing, and maintaining these devices.

3.5 Determining equivalent circuit parameters by test

The actual value of the elements of the transformer equivalent circuit are generally determined by test rather than by calculation. While modern magnetic field

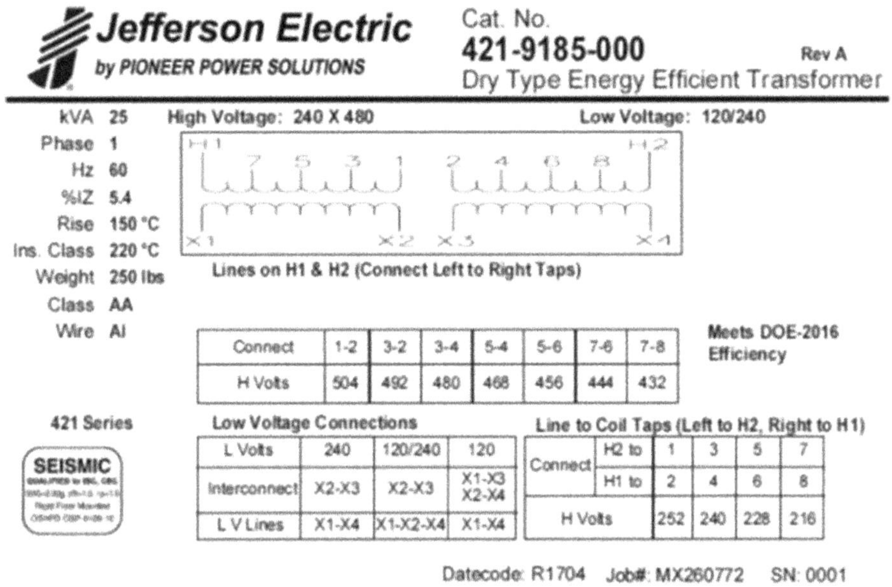

Figure 3.12. Transformer nameplate showing the transformer ratings and connection data. Courtesy of Jefferson Electric.

analytical methods are available which provide good estimates of these values, acceptance tests are generally required for large power transformers. These are performed at the manufacturing facility by the manufacturer, and are generally observed by the buyer. Successful tests are an indication of both good design and construction. They also provide a benchmark for later comparison, after transportation to the installation site, and during subsequent operation.

The two tests of particular interest are the short circuit test and the open circuit test. These test results can be used to determine the transformer equivalent circuit impedance values. These two tests are only one component of the full range of acceptance tests that may be required.

The key to the short circuit and the open circuit tests are that they be conducted at the appropriate rated point—the short circuit test at rated current and temperature, and the open circuit test at rated voltage. Both are conducted at rated frequency. These test parameters ensure that the resulting impedance values will be appropriate for use in the linear model of the transformer.

Open circuit test: The open circuit test is conducted at rated voltage and frequency. It is generally most convenient to apply this voltage to the transformer low side winding. While the test can be done from the high side, the high side power supply will generally be much more expensive for a large power transformer. Note that the low voltage winding can be either the primary or the secondary, depending on the intended use of the transformer. For example, in a generating station, the primary winding will be the low voltage winding, as the power flows into this winding from the generator. This transformer will step the voltage up to transmission voltage levels, in order to achieve good efficiencies in transporting the electric power to the system loads.

The resulting connection for the open circuit test is shown in figure 3.13. The RMS voltage and RMS current and the average real power are measured in this test. Figure 3.13 shows the transformer equivalent circuit with all elements referred to the low voltage side. The high voltage winding is open circuited, so that the only current flow is to magnetize the core. The impedances $R_{cs} = \frac{R_{cp}}{a^2}$ and $X_{ms} = \frac{X_{mp}}{a^2}$ are determined from this test.

With the high side winding open, the equivalent circuit components R_p and X_p will carry no current, and can be ignored in this test. The low side winding impedances R_s and X_s will carry current. These impedances are generally significantly less than the shunt components R_{cs} and X_{ms}, however, and can be neglected. It is a good idea to check that this assumption is valid once the full equivalent circuit parameters have been determined.

The resulting reduced circuit for this test is shown in figure 3.14. The rated value of secondary voltage V_{oc} is applied, and the magnitude of the current I_{oc} and the value of the real power P_{oc} are recorded. Note that figure 3.14 shows phasor values for voltage and current. The angles of these phasors can be inferred from these measured values.

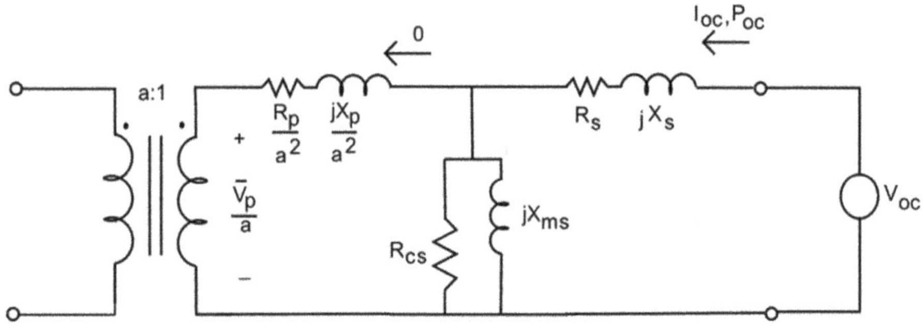

Figure 3.13. Open circuit test schematic.

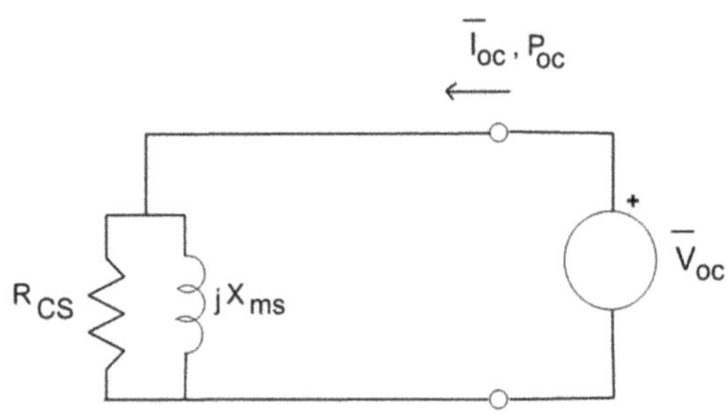

Figure 3.14. Reduced equivalent circuit, open circuit test.

This circuit can be solved in a variety of different ways. One method is as follows:

1. Determine the test volt-amps $S_{oc} = V_{oc} \cdot I_{oc}$
2. Determine the test VARs $Q_{oc} = \sqrt{S_{oc}^2 - P_{oc}^2}$
3. Determine the core loss resistance $R_{cs} = \frac{V_{oc}^2}{P_{oc}}$
4. Determine the magnetizing reactance $X_{ms} = \frac{V_{oc}^2}{Q_{oc}}$
5. When needed, determine the primary equivalents of core loss resistance and magnetizing reactance, by multiplying R_{cs} and X_{ms} by the turns ratio squared.

These steps are for the open circuit test being applied to the secondary of the transformer, which is the preferred method for conducting this test. They would need to be adjusted appropriately if the test were applied to the primary.

Short circuit test: The short circuit test is conducted at rated current. The rated current in the high voltage winding is lower than in the low voltage winding, so the short circuit test is usually applied to the high voltage winding. The low voltage winding is then short circuited. As a result, the voltage level during this test is much less than rated voltage. The magnetizing current during the test is small, and can be neglected for this test.

The resulting simplified equivalent circuit for the short circuit test is shown in figure 3.15. The low side leakage reactance and coil resistance have been referred to the high side winding, and in front of the magnetizing branch. The short circuit on the low voltage side will mean that there is also zero voltage on the high voltage side of the ideal transformer. The resulting reduced equivalent circuit for this test is shown in figure 3.16.

Again, RMS voltage, RMS current, and average power are measured in the test. There are multiple ways to solve this circuit. One method is:

1. Calculate the equivalent series impedance $Z_{sc} = \frac{V_{sc}}{I_{sc}}$
2. Calculate the series resistance $R_{eq} = \frac{P_{sc}}{I_{sc}^2}$
3. Calculate the total leakage reactance $X_{eq} = \sqrt{Z_{sc}^2 - R_{sc}^2}$

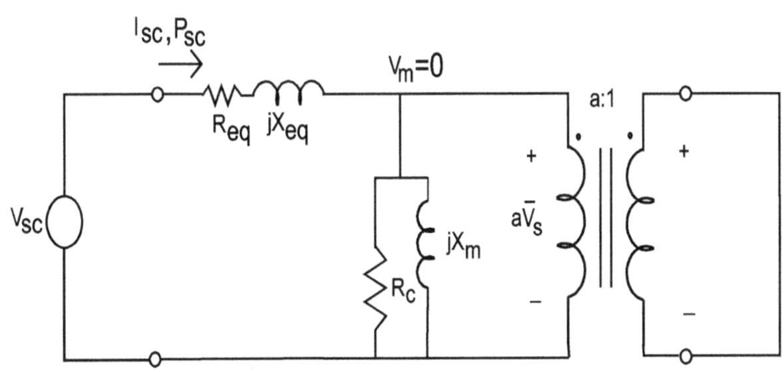

Figure 3.15. Schematic circuit of the short circuit test.

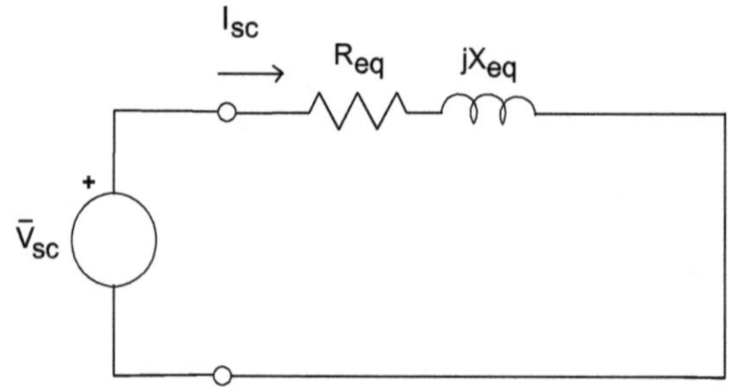

Figure 3.16. Reduced equivalent circuit, short circuit test.

4. The short circuit test is not able to distinguish between the high and low side quantities. Several of the approximate equivalent circuits in figure 3.11 use R_{eq} and X_{eq} in the solution, and work well in many situations. To model the individual coil resistance and leakage reactance components, the following is generally assumed:
$$R_p = a^2R_s = \frac{R_{eq}}{2}$$
$$X_p = a^2X_s = \frac{X_{eq}}{2}$$

These steps assume that the short circuit test is applied to the primary winding. If it is applied to the secondary winding, these formulas need to be adjusted accordingly.

Example 3.7. The transformer of example 3.6 has the following open and short circuit test results:

Test	Short circuit	Open circuit
Excitation applied to	Primary	Secondary
Applied voltage	274 V	240 V
Test current	167 A	175 A
Test real power	12.0 kW	13.33 kW

Find the complete equivalent circuit of this transformer with all values referred to the primary winding.

(a) Short circuit test:

1) $Z_{sc} = \frac{V_{sc}}{I_{sc}} = \frac{274\ \text{V}}{167\ \text{A}} = 1.642\ \Omega$

2) $R_{eq} = \frac{P_{sc}}{I_{sc}^2} = \frac{12.0\ \text{kW}}{167\ \text{A}^2} = 0.432\ \Omega$

3) $X_{eq} = \sqrt{Z_{sc}^2 - R_{sc}^2} = \sqrt{1.642^2 - 0.432^2} = 1.58\ \Omega$

4)
$$R_p = a^2R_s = \frac{R_{eq}}{2} = 0.216\ \Omega$$
$$X_p = a^2X_s = \frac{X_{eq}}{2} = 0.79\ \Omega$$

(b) Open circuit test:

1) $S_{oc} = V_{oc} \cdot I_{oc} = 240\text{ V} \cdot 175\text{ A} = 42.16\text{ kVA}$

2) $Q_{oc} = \sqrt{S_{oc}^2 - P_{oc}^2} = \sqrt{42.16\text{ kVA}^2 - 13.33\text{ kW}^2} = 40.0\text{ kVAR}$

3) $R_{cs} = \frac{V_{oc}^2}{P_{oc}} = \frac{240\text{ V}^2}{13.33\text{ kW}} = 4.32\ \Omega$

4) $X_{ms} = \frac{V_{oc}^2}{Q_{oc}} = \frac{240\text{ V}^2}{40.0\text{ kVAR}} = 1.44\ \Omega$

5) The corresponding primary values of core loss resistance and magnetizing reactance are then

$$R_{cp} = 20^2 \cdot 4.32\ \Omega = 1730\ \Omega$$
$$X_{mp} = 20^2 \cdot 1.44\ \Omega = 576\ \Omega$$

The resulting equivalent circuit is shown in figure 3.17, with all parameters referred to the hi side.

3.6 Power transformer thermal model

The steady state volt-amp rating of power transformers is generally determined by its thermal performance and insulation temperature ratings. Over the long term, insulation temperature is a primary factor in aging and ultimately in the service life of the transformer.

There are a variety of insulation materials used in transformer windings. These have been classified by both NEMA (National Electric Manufacturers Association) and IEC (International Electrotechnical Commission). Several insulation classes and example insulating materials are shown in table 3.1.

Power transformers are generally cooled with transformer oil. The core and winding assembly are immersed in oil within a tank. An example distribution transformer in a tank is shown in figure 3.1(a). Larger transformers have radiators to help disperse the heat generated in the core and winding, and some of these include fans to blow air over the radiators and perhaps pumps to circulate the oil. The tanks

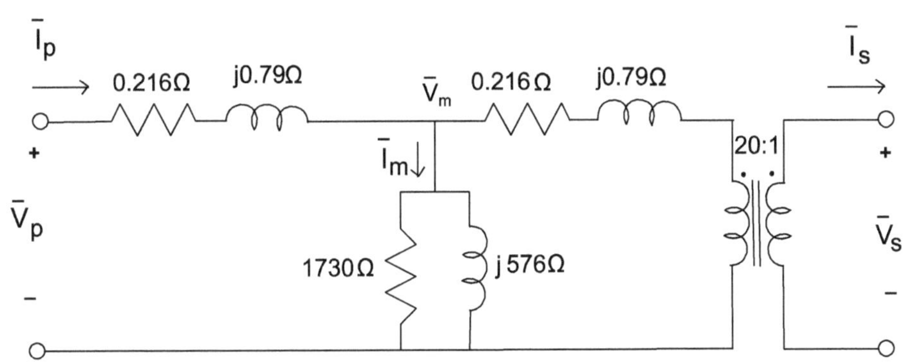

Figure 3.17. Transformer equivalent circuit.

are generally sealed to minimize the moisture content in the oil, and typically have an inert gas blanket to accommodate thermal expansion. Oil circulating through the windings and around the core captures heat, and moves it to areas where it can be dissipated to the atmosphere.

Power transformers are rated for a maximum ambient temperature for which they can operate at full load. Under load, the transformer oil temperature will increase. Oil cooled transformers typically have a temperature sensor mounted at the top of the tank—the measured value is referred to as the top oil temperature. In addition to this, the conductor temperature will be higher than the top oil temperature—the hottest temperature in the conductor is known as the hot spot temperature. This hot spot temperature needs to be lower than the thermal rating of the class of insulation used in the winding.

A conceptual cross sectional diagram of a large power transformer is shown in figure 3.18. Thermal energy is created in the core and coils of the transformer. This raises the temperature of these components, and this energy will be transferred to the transformer oil. Through convection, the hot oil will rise through the coil/core structure, and will circulate through the radiators.

Table 3.1. NEMA and IEC thermal insulation classes.

NEMA letter insulation class	IEC 60085 thermal class	Maximum hot spot temperature	Typical material
A	105	105 °C	Cotton, paper, etc
B	130	130 °C	Inorganics such as mica, glass fiber
	120	120 °C	Polyurethane, etc
F	155	155 °C	Inorganics with high temp binders
H	180	180 °C	Silicone elastomers, etc

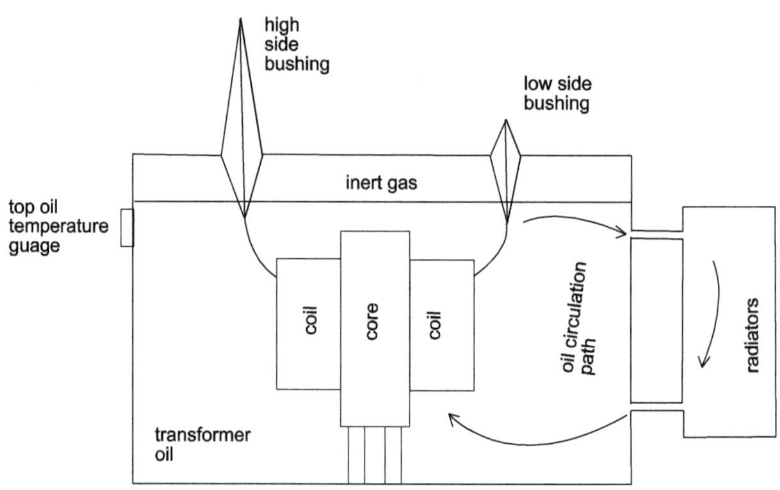

Figure 3.18. Cross section diagram of a large power transformer.

Through this mechanism, the highest temperatures will be in the core and coil. The oil temperature will increase as it raises through the core/coil structure, and then will decrease as it circulates through the radiators. The temperature across the core and coil is not constant, and the designers will identify a temperature gradient in this structure. Typically, the coil temperature is the first concern. The hottest temperature in the coil is called the 'hot spot' temperature, and this temperature must stay within the temperature ratings of the insulation of the coil.

ANSI/IEEE Standard C57.91-1995 is titled 'IEEE Guide for Loading Mineral Oil Immersed Power Transformers up to 100 MVA with 55 °C or 65 °C Average Winding Rise.' This standard provides a method for predicting the hot spot temperature of a transformer. The standard defines the following terms:

θ_a = ambient temperature
θ_g = hot spot conductor rise over top oil temperature
$\theta_{g(fl)}$ = hot spot conductor rise over top oil temperature at full load

θ_{hs} = actual conductor hot spot temperature
θ_o = top oil rise over ambient temperature
θ_{fl} = top oil rise over ambient at full load

R = ratio of load loss to no load loss at rated load
K = ratio of load L to rated load

For constant loads applied over the long term, the hot spot temperature will be

$$\theta_{hs} = \theta_a + \theta_o + \theta_g$$

The rise in the top oil temperature over ambient temperature will be

$$\theta_o = \theta_{fl}\left[\frac{K^2R + 1}{R + 1}\right]^{0.9}$$

The rise in the hot spot temperature over top oil temperature will be

$$\theta_g = \theta_{g(fl)}\,K^{1.6}$$

The values of θ_{fl} and $\theta_{g(fl)}$ are known for the case when the transformer is operating at rated load and rated voltage at the rated ambient temperature. The value of R can be determined from the transformer equivalent circuit values. With this knowledge, the top oil and winding hot spot temperatures can be predicted for a given operating condition.

Example 3.8. An oil cooled power transformer has a rated load of 40 MVA. This rating is valid at an ambient temperature of 40 °C. It has NEMA Class B insulation, and it is designed to operate with a maximum hot spot temperature of 120 °C, and a top oil temperature of 75 °C at its rated point. The ratio of load loss to no load loss at rated load is $R = 5.0$. Determine the transformer top oil and hot spot temperatures when the transformer is operated at half rated load with and ambient temperature of 30 °C.

From the data at the rated point,

$$\theta_{\text{fl}} = 75^\circ - 40^\circ = 35^\circ$$
$$\theta_{\text{g(fl)}} = 120^\circ - 75^\circ = 45^\circ$$

At half load,

$$K = \frac{0.5R}{R} = 0.5.$$

The top oil temperature rise above ambient will be

$$\theta_o = 35\left[\frac{0.5^2 5.0 + 1}{5.0 + 1}\right]^{0.9} = 14.5\ ^\circ\text{C}$$

The hot spot temperature rise above top oil will be

$$\theta_{\text{g}} = 45(0.5)^{1.6} = 14.9\ ^\circ\text{C}$$

The top oil temperature will then be

$$\theta_{\text{top oil}} = \theta_{\text{a}} + \theta_o = 30 + 14.5 = 44.5\ ^\circ\text{C}$$

The hot spot temperature will be

$$\theta_{\text{hs}} = 30 + 14.5 + 14.9 = 59.4\ ^\circ\text{C}$$

This is significantly cooler than at the rated point. Some transformer designs employ smaller radiators with fans forcing air past the radiator. In these designs, the fans will be run only when needed. Transformer ratings are given for both fans running and fans off, and the thermal constants for the transformer must be used for the correct operating situation.

Example 3.9. For this same transformer, calculate the amount of load L that would be required to produce a 120 °C hot spot temperature when the ambient temperature is −20 °C.

From the data at the rated point,

$$\theta_{\text{fl}} = 75^\circ - 40^\circ = 35^\circ$$
$$\theta_{\text{g(fl)}} = 120^\circ - 75^\circ = 45^\circ$$

At −20 °C, the top oil temperature rise above ambient will be

$$\theta_o = 35\left[\frac{K^2 \cdot 5.0 + 1}{5.0 + 1}\right]^{0.9}$$

The hot spot temperature rise above top oil temperature will be

$$\theta_g = 45 \cdot K^{1.6}$$

Finally, the actual hot spot temperature will be

$$\theta_{hs} = \theta_a + \theta_o + \theta_g$$

or

$$120 = -20 + \theta_o + \theta_g$$

This non-linear set of equations must be solved by iteration. One way to do this is to set up a spreadsheet with K as a variable, and then evaluate the hot spot temperature equation to determine if the right hand side of this equation does equal 120°. Using spreadsheet tools, it is straightforward to increment K by a series of small values until this equality is true.

By doing this, a value of $K = 2.22$ is found. In other words, this transformer can be loaded to more than double its rated load when the ambient temperature is −20 °C, without exceeding the rated transformer winding temperature. Note that there may be limits on other transformer components that would be overstressed at this loading—the transformer manufacturer would need to be consulted to determine if this is the case.

3.7 Frequency response of signal transformers

Transformers are used in a variety of instrumentation and signal applications. For example, audio amplifiers regularly include transformers. In audio applications, these transformers are required to operate appropriately over a wide frequency range.

At low frequencies, the leakage inductances and the core loss resistance are not important. The transformer model shown in figure 3.19 can be used to determine the low cutoff frequency, where the gain V_{out}/V_{in} drops 3 db below its center frequency gain.

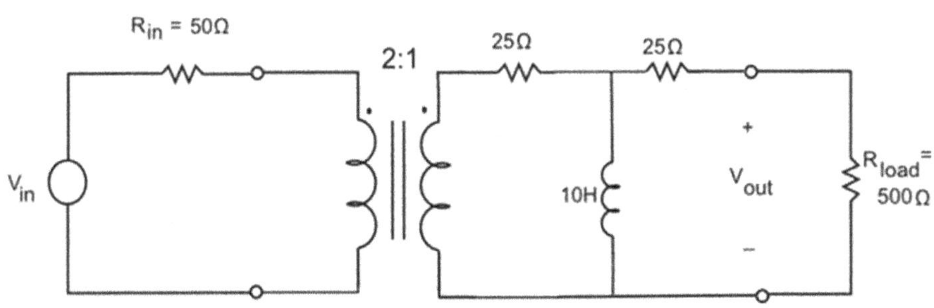

Figure 3.19. Low frequency model of an audio transformer with representative parameters.

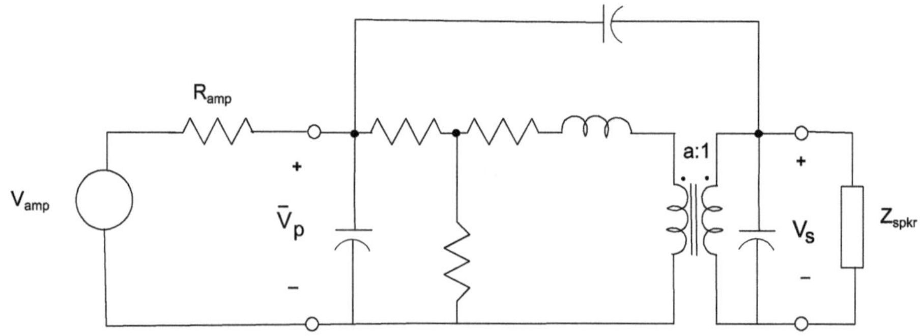

Figure 3.20. High frequency model of audio transformer.

The high frequency performance of the transformer is also of interest. At high frequencies, the performance is impacted by the stray capacitances of the transformer. An approximate high frequency equivalent circuit of the transformer is shown in figure 3.20. This equivalent circuit accounts for the stray capacitance by including a shunt capacitor across both the primary and secondary windings, and a series capacitance between the primary and secondary windings. The model also includes the coil and core resistances, and the primary and secondary leakage inductances, which are lumped together in the primary side of the model. This model can be used with the source Thevenin equivalent and the load to determine the high frequency performance of an audio transformer.

Audio transformers can provide impedance matching, signal amplification, and other advantages. With modern electronics, however, these can also be provided by electronic circuits. Transformers do provide galvanic isolation to reduce ground loops, improve safety and protect equipment. Audio transformers do impact the system audio output, and some prefer the use of audio transformers for this reason. More information on the topic of audio transformers can be found in [1].

Reference

[1] Whitlock B 2001 Audio transformers *Handbook for Sound Engineers* 3rd ed G Ballou (Waltham, MA: Focal Press) ch 11

Questions

1. Find a manufacturer of pole top distribution transformers that has product information on the web. From the product literature, record the main product ratings.
2. Figure 3.12 shows the nameplate of a dry type transformer. On the web, find the specification sheet and/or product description for this or a similar line of transformers. Identify information that is on the nameplate but not on the spec sheet or product description. Also, identify information that is on the spec sheet/product description but not on the nameplate.

Problems

1. A high impedance microphone has an output impedance of 10 000 ohms. It is feeding an amplifier with an input impedance of 2000 ohms.
   a. If the microphone open circuit voltage is 50 mV, what would be the voltage at the input to the amplifier? What is the power delivered to the amplifier?
   b. An impedance matching transformer is being considered to increase the power to the amplifier. Using the ideal transformer model, choose an isolation transformer ratio that would make the apparent impedance of the microphone to be 2000 ohms as viewed from the amplifier. What is the power delivered to the amplifier when this transformer is in inserted between the microphone output and the amplifier input.
2. Power distribution transformers that feed residences usually have two identical secondary windings, with each winding having a turns ratio of 60 with respect to the primary. As shown in the figure below, the two windings are connected in series, and the neutral connection point grounded. 120 V loads are connected between the 'hot' leads and the neutral. 240 V loads are connected between the two hot leads. Use the ideal transformer model to calculate the four currents $\bar{I}_p$, $\bar{I}_{s1}$, $\bar{I}_{s2}$, and $\bar{I}_n$ for the following cases:
   a. $R_1 = 5\ \Omega$ and $R_2 = 5\ \Omega$
   b. $R_1 = 10\Omega$ and $R_2 = 5\Omega$

(Note the polarity marks for the coils. If your neutral current is larger than the line currents, you have your polarities wrong.)

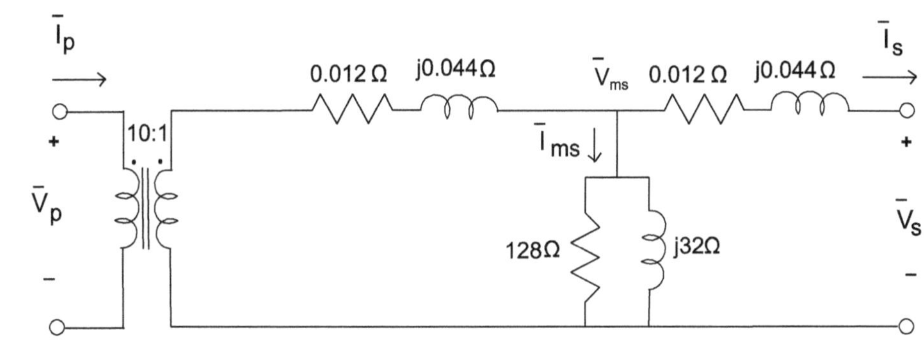

3. An ideal transformer has a turns ratio of 8. It is feeding a purely resistive load of 20 ohms.
   a. If there is an AC current of 2 A rms going through the load, calculate the voltage at the transformer secondary. Assume an angle of 0 degrees for this current.
   b. Calculate the primary voltage and the primary current going into the transformer.
   c. Calculate the ratio of the primary voltage to primary current. (This is the effective resistance of the load as seen from the primary.)

d. What is the relationship between the answer in part c and the turns ratio?

e. Calculate the real power flowing into the transformer and the real power flowing out of the transformer.

4. A 60 Hz transformer has a rating of 250 kVA, 2400 V:120 V. Determine the transformer turns ratio, and the rated primary and secondary currents.
5. A 60 Hz transformer has the following parameters:

   Turns ratio: 40
   Primary coil resistance $R_p = 2.4$ ohms
   Primary leakage reactance $X_p = 9.5$ ohms
   Magnetizing reactance $X_{mp} = 8500$ ohms
   Core loss resistance $R_{cp} = 14\,500$ ohms
   Secondary coil resistance $R_s = 0.0015$ ohms
   Secondary leakage reactance $X_s = 0.006$ ohms

   The transformer secondary voltage is 120 V, and it is delivering 25 A to a load. The load current lags the voltage by 25°. Use a transformer model of figure 3.9 in this problem.

   a) Draw the transformer equivalent and label the impedance values.
   b) Find the primary voltage and current.
   c) Find the input and output real powers of the transformer.
   d) Find the transformer efficiency.
6. Repeat the previous problem, using the equivalent circuit of figure 3.11(a). Discuss the reduction in accuracy of this model as compared to the figure 3.9 model.
7. A single phase 60 Hz, 500 kVA 2400 V:480 V transformer has the following data from short circuit and open circuit testing

Open circuit test	Winding	Quantity	Value
	Secondary	V_{oc}	480 V
		I_{oc}	49 A
		P_{oc}	6.67 kW
	Primary	Open circuited	
Short circuit test	Primary	V_{sc}	129 V
		I_{sc}	208 A
		P_{sc}	10.0 kW
	Secondary	Short circuited	

   Using this test data, determine the values of the transformer turns ratio and impedances, and draw the full equivalent circuit with these values shown on the circuit.
8. A 300 kVA power transformer is rated to have a top oil temperature rise above ambient of $\theta_{fl} = 35$ °C at rated load. Also at rated load, the hot spot conductor rise over top oil temperature is $\theta_{g(fl)} = 30$ °C. The transformer designed to serve this load at ambient temperatures of up to $\theta_a = 40$ °C,

when the actual conductor hot spot temperature will be $\theta_{hs} = 105$ °C. Under rated conditions, the transformer has a no load loss of 4 kW, and a load loss of 6 kW.

Determine the transformer top oil temperature when the transformer is operating at a load of 150 kVA at an ambient temperature of 50 °C.

For the same transformer, if the ambient temperature will go no higher than 5 °C on a given day, what load can be placed on this transformer so that the predicted hot spot temperature will not exceed $\theta_{hs} = 105$ °C?

IOP Publishing

Thomas Ortmeyer

Chapter 4

Three phase transformer banks

Three phase transformer banks are widely used in the AC power grid. Generators of size above a few kilowatts connect to the grid with three phase, and power grid customers apart from residential and some small commercial are served with three phase. Three phase transformer banks are used to step the generator voltage up to transmission or distribution levels, and then step it back down from transmission to distribution and from primary distribution to secondary distribution and utilization levels. Transmission level voltages are required to move the power from generation to load centers and between regions, while primary distribution level voltages are necessary to move power effectively within communities. Utilization level voltages—for example, 120 V—are typically only used to move power for tens of meters.

4.1 Three phase transformer cores

Three phase transformers can be constructed on a single core, or can be assembled from three individual single phase transformers.

Figure 4.1 shows the diagram of a three phase transformer wound on a single core, in the core form style. There is also a shell form style for three phase transformers. The three legged core form transformer has some unique properties during unbalanced operation.

In this diagram, the primary phase windings are designated with the upper case variables A, B and C. The secondary phase windings are designated with the lower case variables a, b, and c. This notation is used throughout this chapter. In this diagram, the primary winding has more turns than the secondary winding—and so has a higher voltage rating. This transformer is then intended to be used as a step down transformer. Note that the transformer itself doesn't care which direction power flows—the designation of primary versus secondary is strictly for the convenience of the user. Often, it can be more convenient to refer to the high voltage (sometimes 'hi') winding and the low voltage ('lo') winding, as the terms

doi:10.1088/978-0-7503-1662-0ch4

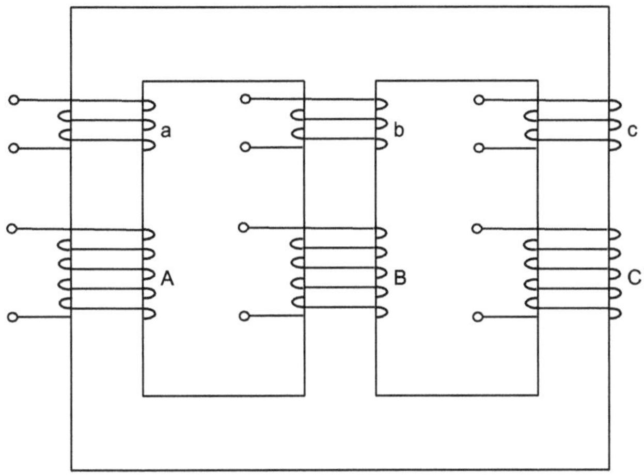

Figure 4.1. Three phase core form transformer, showing primary (A, B, C) and secondary (a, b, c) windings.

'primary' and 'secondary' can be ambiguous as it depends on the particular application.

A three phase transformer bank can also be constructed from three individual transformers, each with a separate core as discussed in chapter 3. Strictly speaking, the term 'transformer bank' refers to this connection of three separate transformers to make a single three phase transformer.

Also note in figure 4.1 that the high and low windings are wound in the same direction. By the right hand rule, the polarity marks would be on the same side (top or bottom) of the high and low windings on each individual leg of the transformer.

Three phase transformers will be symmetrical—the individual phase coils of the high side winding will be identical, each have the same number of turns and the same wire size and construction. They will have the same resistance, leakage reactance, and current carrying capability. Similarly, the low side winding phase coils will be identical to each other. Further, the high and low side windings will be designed for compatibility of their respective current carrying capability, as well as to provide the desired voltage ratio.

4.2 Three phase transformer windings

The two most common windings for three phase transformers are the wye (or Y) connected and the delta (or star) connected windings. This refers to electrical connection of the three phase windings. Both the primary and secondary winding of the transformer shown in figure 4.1 can be connected in either wye or delta.

4.2.1 Wye winding

In the wye winding, the individual phase coils that make up the winding are connected from line to neutral, as shown in figure 4.2. The neutral point can either be solidly grounded, grounded through an impedance, or left unconnected. During perfectly balanced operation of the three phase system, there will be negligible current flow

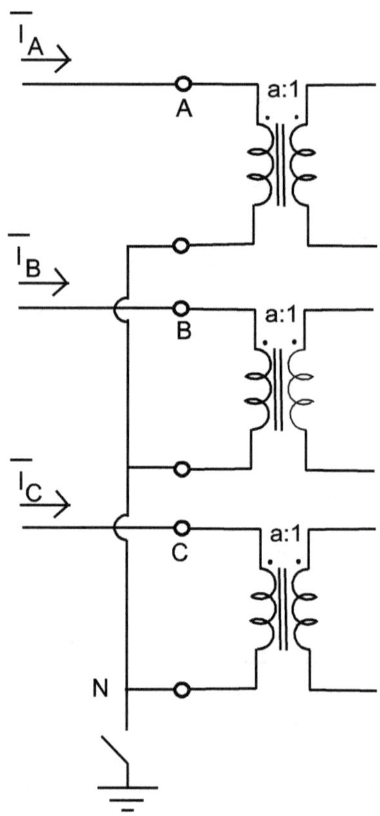

Figure 4.2. Three phase transformer with primary winding connected in wye.

between neutral and ground when the switch is closed. When the system is unbalanced, however, the ground connection or lack of one at the neutral point becomes very important. The grounding of transformers and of the power system in general is considered in detail in advanced courses and studies of power system engineering.

In figure 4.2, the non-polarity side of the primary phase coils are connected together at the neutral (N) point. The figure shows a switch between neutral and ground, to recognize that the neutral may or may not be grounded. The polarity side of each phase is connected to the phases of the power system. In this wye connection, the line current coming in on each phase will go directly into the transformer coil of that phase. The phase windings are connected from the line to neutral, so the voltage across each coil will be the line to neutral voltage. When the neutral point is grounded, the phase voltages will also be line to ground.

A word about notation—the open circles in this diagram show the transformer terminals—these define the boundary between transformer and the incoming line. The semi-circles in the neutral connection line show that there is no connection between this neutral line and the phase connection that it passes over. In this notation, when the wires come together without this semi-circle, then there is a connection at that point.

Figure 1.6 shows the relationship between the line to neutral and the line to line voltages for the wye winding—these relationships are the same as for any wye connected load. The result is that the magnitude of the line to line voltage is $\sqrt{3}$ times the magnitude of the line to neutral voltage, and there is a 30° shift between the line to neutral and line to line voltages.

4.2.2 Delta connection

A delta connection is shown in figure 4.3. In a delta connection, the non-polarity side of one phase coil is connected to the polarity side of the next coil. In figure 4.3, non-polarity of the A phase is connected to polarity of the B phase, and so on. In the delta connection, the phase coils are connected line to line. In figure 4.3, the winding currents within the transformer windings are given the subscripts 1, 2, and 3, while the line currents coming into the transformer are given the A, B and C subscripts.

With balanced currents in the winding, the line currents can be found. The phasor relationship between winding and line currents in this delta winding is shown in figure 4.4.

By examining figure 4.3, it can be determined that the A phase line current is

$$\overline{I}_A = \overline{I}_1 - \overline{I}_3 \tag{4.1}$$

Similar relationships exist between $\overline{I}_B$ and $\overline{I}_2$ and $\overline{I}_1$, and $\overline{I}_C$ and $\overline{I}_3$ and $\overline{I}_2$.

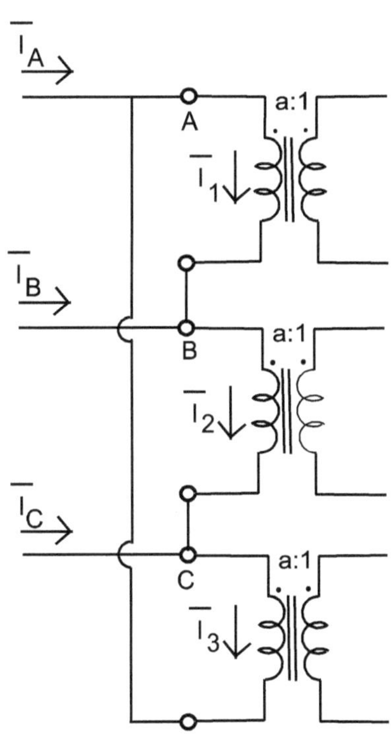

Figure 4.3. Delta connection of the primary winding.

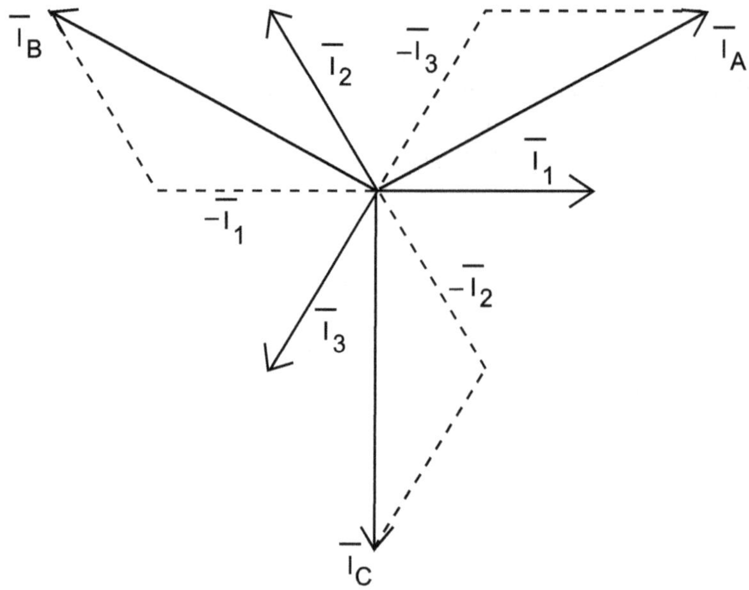

Figure 4.4. The relationship between winding currents and line currents in the delta winding of figure 4.3.

This is shown graphically in figure 4.4. The result for the three line currents in the balanced case is

$$\begin{aligned} \overline{I}_A &= \sqrt{3} \cdot \overline{I}_1 \cdot 1 \underline{/30^\circ} \\ \overline{I}_B &= \sqrt{3} \cdot \overline{I}_2 \cdot 1 \underline{/30^\circ} \\ \overline{I}_C &= \sqrt{3} \cdot \overline{I}_3 \cdot 1 \underline{/30^\circ} \end{aligned} \quad (4.2)$$

It is worth noting that this is one of two ways to connect the delta winding. In the alternate connection, the non-polarity side of the A phase is connected to the polarity side of C phase, so that

$$\overline{I}_A = \overline{I}_1 - \overline{I}_3 \quad (4.1)$$

The other two phases are connected accordingly, so that $\overline{I}_B = \overline{I}_2 - \overline{I}_1$ and $\overline{I}_C = \overline{I}_3 - \overline{I}_2$. This results in the same magnitude difference but a somewhat different phase angle relationship between line and winding currents.

As a result of the use of wye and delta windings, three basic types of three phase transformers exist—

- wye–wye,
- delta–wye,
- delta–delta.

Some would say that the wye–delta (wye primary and delta secondary) is a fourth type of transformer, but we do not make that distinction in this book. Also, note that there can be other types of windings—but the delta and wye are the most prominent by far. Also note that this chapter discusses two winding transformers. There are also three winding transformers in common usage. Finally, autotransformers are commonly used as well, particularly in transmission applications. These types of transformers are covered in advanced courses.

The initial analysis of these transformers is done using the ideal transformer model. The effects of transformer winding resistance, leakage and magnetizing reactances, and core loss will be considered in a later section.

4.3 Wye–wye transformers

The connection diagram of a wye–wye transformer is shown in figure 4.5. Each phase on both the primary and secondary windings is connected from line to neutral. Figure 4.5 shows both neutrals connected to directly to ground. In practice, one or both neutrals could be floating, or could be connected to ground through a resistor or reactor. Under balanced operating conditions, there is no neutral to ground current, and all of these connections are analysed in the same way.

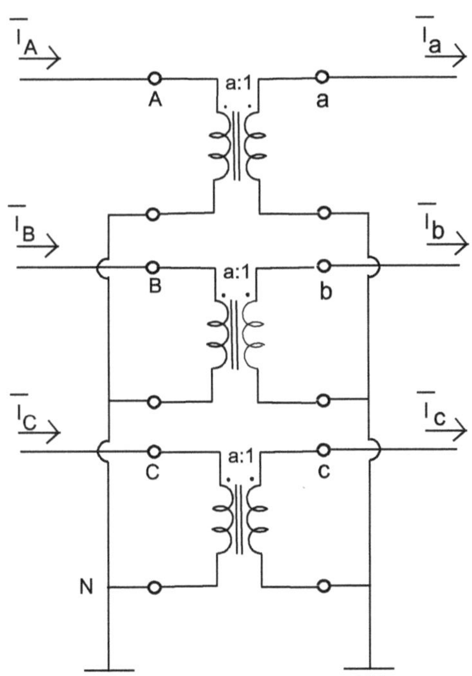

Figure 4.5. Wye–wye connected transformer.

Consider the transformer using the ideal transformer theory for each of the three transformers that make up the bank. From the figure, the relationship between the primary and secondary phase currents is:

$$\bar{I}_{\mathrm{A}} = \frac{\bar{I}_{\mathrm{a}}}{a} \qquad \bar{I}_{\mathrm{B}} = \frac{\bar{I}_{\mathrm{b}}}{a} \qquad \bar{I}_{\mathrm{C}} = \frac{\bar{I}_{\mathrm{c}}}{a} \tag{4.4}$$

It is clear from equation (4.4) that the primary currents will be in phase with the secondary currents, but reduced by the turns ratio a. The resulting phasor diagram for the currents is shown in figure 4.6.

With the phase windings connected between line and ground, the transformer winding voltages equal the line to ground voltages,

$$\bar{V}_{\mathrm{A}} = a\bar{V}_{\mathrm{a}} \qquad \bar{V}_{\mathrm{B}} = a\bar{V}_{\mathrm{b}} \qquad \bar{V}_{\mathrm{C}} = a\bar{V}_{\mathrm{c}} \tag{4.5}$$

In the balanced case, the secondary line to ground voltages will have equal magnitude and be separated in time by 120°

$$\bar{V}_{\mathrm{a}} = V_{\mathrm{s}}\underline{/\delta_{\mathrm{s}}} \qquad \bar{V}_{\mathrm{b}} = V_{\mathrm{s}}\underline{/\delta_{\mathrm{s}} - 120^{\circ}} \qquad \bar{V}_{\mathrm{c}} = V_{\mathrm{s}}\underline{/\delta_{\mathrm{s}} - 240^{\circ}} \tag{4.6}$$

From equations (4.5) and (4.6), the primary phase to neutral voltages will be in phase with the secondary phase to neutral voltages, but with a magnitude of aV_{s}.

On both primary and secondary, the line to line voltages will be larger than their respective line to ground voltage by the factor $\sqrt{3}$, as shown in figure 1.6. The line to

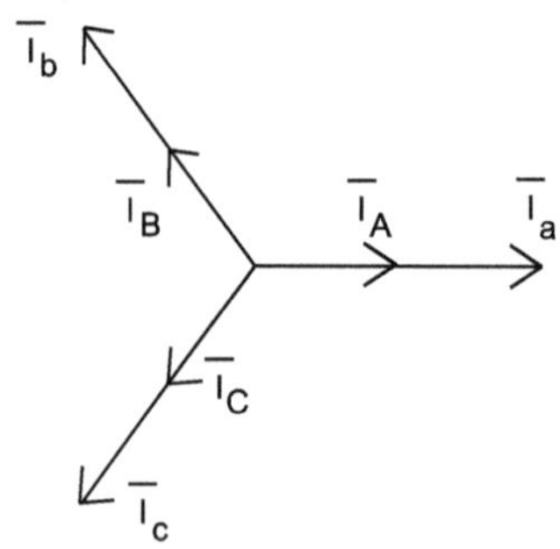

Figure 4.6. Primary and secondary current phasor diagram for the wye–wye transformer.

line voltages will have a 30° between line to line and line to neutral voltages. This phase shift can be either leading or lagging, depending on the definition of the set of line to line voltages as either $\overline{V}_{ab}$, $\overline{V}_{bc}$, $\overline{V}_{ca}$ or $\overline{V}_{ac}$, $\overline{V}_{ba}$, $\overline{V}_{cb}$. Both options can be found in practice.

4.4 Delta–wye transformers

The connection diagram of a delta–wye transformer is shown in figure 4.7.

Figure 4.7 shows that the secondary currents are reflected onto the primary currents within the delta winding by the winding turns ratio a. For the case where the A phase current is at phase angle ϕ, the following relationships hold for the balanced case,

$$\begin{aligned}\overline{I}_1 &= \frac{\overline{I}_a}{a} = \frac{I_a \underline{/\phi^\circ}}{a}\\ \overline{I}_2 &= \frac{\overline{I}_b}{a} = \frac{I_a \underline{/\phi - 120^\circ}}{a}\\ \overline{I}_3 &= \frac{\overline{I}_c}{a} = \frac{I_a \underline{/\phi + 120^\circ}}{a}\end{aligned} \tag{4.7}$$

The primary line currents $\overline{I}_A$, $\overline{I}_B$ and $\overline{I}_C$ are then determined from the winding currents by Kirchoff's current law, as shown in figure 4.8. From equations (4.2) and (4.7), the primary line currents are

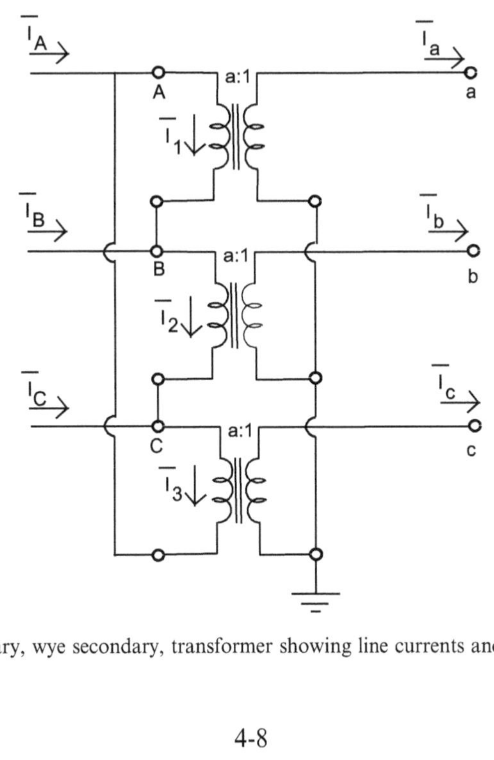

Figure 4.7. Delta primary, wye secondary, transformer showing line currents and delta winding currents.

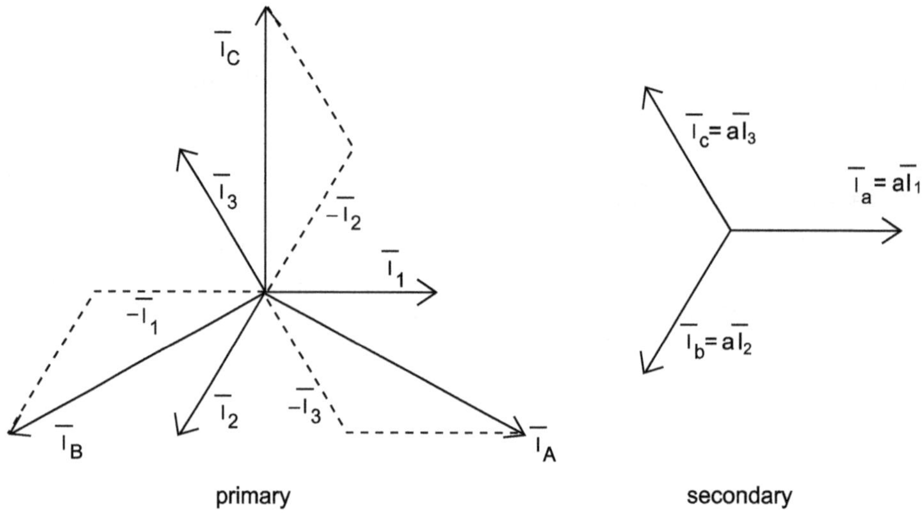

Figure 4.8. Primary and secondary phasor currents for the delta–wye transformer of figure 4.7.

$$\begin{aligned}
\bar{I}_A &= \bar{I}_1 - \bar{I}_3 = \sqrt{3}\cdot\frac{\bar{I}_a}{a}\cdot 1\,\underline{/30^\circ} = \frac{\sqrt{3}}{a}I_a\,\underline{/\phi - 30^\circ}\\
\bar{I}_B &= \bar{I}_2 - \bar{I}_1 = \sqrt{3}\cdot\frac{\bar{I}_b}{a}\cdot 1\,\underline{/30^\circ} = \frac{\sqrt{3}}{a}I_a\,\underline{/\phi - 150^\circ}\\
\bar{I}_C &= \bar{I}_3 - \bar{I}_2 = \sqrt{3}\cdot\frac{\bar{I}_c}{a}\cdot 1\,\underline{/30^\circ} = \frac{\sqrt{3}}{a}I_a\,\underline{/\phi + 90^\circ}
\end{aligned} \tag{4.8}$$

The phasor diagram for these three phase currents is shown in figure 4.8. In all three phases, the primary line current is smaller than the secondary line current by a factor of $\frac{a}{\sqrt{3}}$. Also in each case, the primary line current leads the secondary line current of the same phase by 30°. Note that the apparent ratio between the primary and secondary line currents is $\frac{a}{\sqrt{3}}$. This is only the case under balanced conditions.

In a similar fashion, the line to ground secondary voltages relate to the line to line primary voltages. In the case where the secondary a phase voltage is at 0° and the secondary line to neutral voltage magnitude is V_a,

$$\begin{aligned}
\bar{V}_{AB} &= a\bar{V}_a = aV_a\,\underline{/0^\circ}\\
\bar{V}_{BC} &= a\bar{V}_b = aV_a\,\underline{/-120^\circ}\\
\bar{V}_{CA} &= a\bar{V}_c = aV_a\,\underline{/120^\circ}
\end{aligned} \tag{4.9}$$

In this case, it must be inferred that during balanced operation, the supplying source on the primary side is grounded wye. The line to line secondary voltages will be

$$\overline{V}_{ab} = \overline{V}_a - \overline{V}_b = \sqrt{3}\, V_a \underline{/30^\circ}$$
$$\overline{V}_{bc} = \overline{V}_b - \overline{V}_c = \sqrt{3}\, V_a \underline{/-90^\circ} \qquad (4.10)$$
$$\overline{V}_{ca} = \overline{V}_c - \overline{V}_a = \sqrt{3}\, V_a \underline{/150^\circ}$$

From equations (4.9) and (4.10), the magnitude ratio of the primary line to line voltage and the secondary line to line voltage will be $\frac{a}{\sqrt{3}}$ and the secondary line to line voltage will lead the primary line to line voltage by 30°. These relationships are shown in figure 4.9.

The current phasor diagram of figure 4.8 can be rotated as a whole to account for phase angles other than 0° in the secondary a phase current. Similarly, the voltage phasor diagrams of figure 4.9 can be rotated as appropriate, to account for other angles of the A phase to ground voltage.

Alternate connection of delta–wye transformer

Figure 4.10 shows the alternate connection of the delta winding—with the A coil connected between A and C phases rather than between the A and B phases. Analysis of this version of the delta–wye transformer shows that the ratio between primary and secondary line currents and line to line voltages remains at $\frac{a}{\sqrt{3}}$, but the line quantities are shifted forward by 30° rather than back by 30°.

4.5 Delta–delta transformers

Delta–delta transformers have delta windings on both the primary and secondary. A delta–delta transformer is shown in figure 4.11. As expected, the winding currents and line currents are different in both delta windings. Figure 4.11 shows winding currents $\overline{I}_1$ through $\overline{I}_3$ on the primary and $\overline{I}_4$ through $\overline{I}_6$ on the secondary.

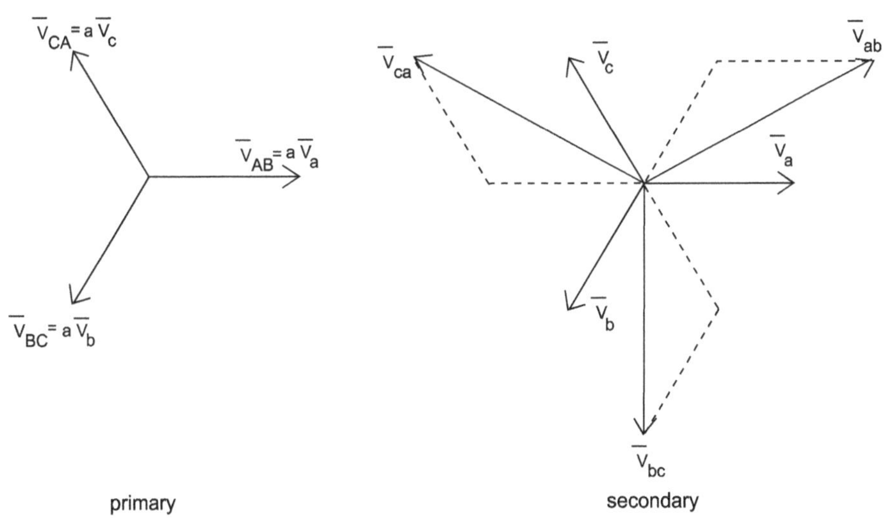

Figure 4.9. The relationship between primary and secondary line to line voltages.

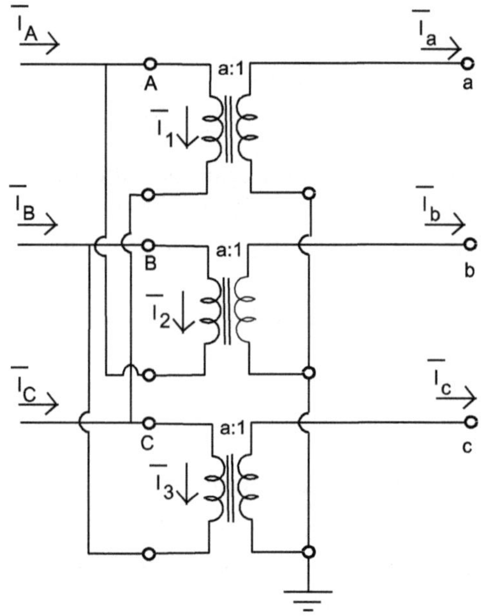

Figure 4.10. Delta–wye transformer with +30° phase shift.

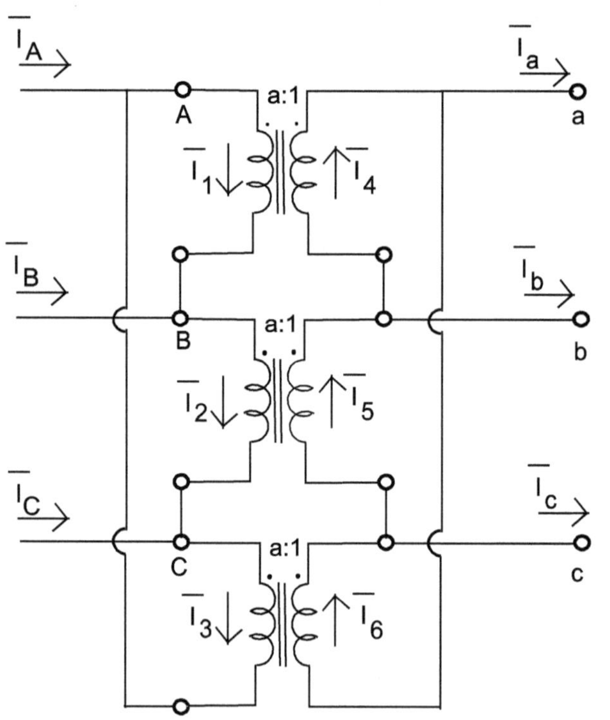

Figure 4.11. Delta–delta transformer connection.

From figure 4.11, the secondary line currents are

$$\begin{aligned}\bar{I}_a &= \bar{I}_4 - \bar{I}_6\\ \bar{I}_b &= \bar{I}_5 - \bar{I}_4\\ \bar{I}_c &= \bar{I}_6 - \bar{I}_5\end{aligned} \tag{4.11}$$

Consider the secondary winding currents to be:

$$\begin{aligned}\bar{I}_4 &= I_4\,\underline{/0^\circ}\\ \bar{I}_5 &= I_4\,\underline{/-120^\circ}\\ \bar{I}_6 &= I_4\,\underline{/120^\circ}\end{aligned} \tag{4.12}$$

Note that I_4 is designated as the current magnitude for the magnitude of the balanced secondary winding currents.

Combining these two equations, the secondary line currents will be

$$\begin{aligned}\bar{I}_a &= \bar{I}_4 - \bar{I}_6 = \sqrt{3}\,I_4\,\underline{/-30^\circ}\\ \bar{I}_b &= \bar{I}_5 - \bar{I}_4 = \sqrt{3}\,I_4\,\underline{/-150^\circ}\\ \bar{I}_c &= \bar{I}_6 - \bar{I}_5 = \sqrt{3}\,I_4\,\underline{/90^\circ}\end{aligned} \tag{4.13}$$

On the primary, the winding currents will be

$$\begin{aligned}\bar{I}_1 &= \frac{1}{a}\bar{I}_4\\ \bar{I}_2 &= \frac{1}{a}\bar{I}_5\\ \bar{I}_3 &= \frac{1}{a}\bar{I}_6\end{aligned} \tag{4.14}$$

For the winding currents of equation (4.12), the primary line currents will be

$$\begin{aligned}\bar{I}_A &= \bar{I}_1 - \bar{I}_3 = \frac{\sqrt{3}}{a}I_4\,\underline{/-30^\circ} = \frac{1}{a}\bar{I}_a\\ \bar{I}_B &= \frac{\sqrt{3}}{a}I_4\,\underline{/-150^\circ} = \frac{1}{a}\bar{I}_b\\ \bar{I}_C &= \frac{\sqrt{3}}{a}I_4\,\underline{/90^\circ} = \frac{1}{a}\bar{I}_c\end{aligned} \tag{4.15}$$

Comparing equations (4.13) and (4.15), it is apparent that the primary and secondary current magnitudes are related by a factor of turns ratio a. The primary and secondary line current angles are equal in the equivalent phases. Note that this is the same relationship as exists between primary and secondary line currents in the wye–wye transformer.

In both sides of the delta–delta transformer, the windings are connected line to line, so the winding voltages equal the line to line voltages. The relationship between the primary and secondary line to line voltages is then

$$\begin{aligned} \overline{V}_{AB} &= a\overline{V}_{ab} \\ \overline{V}_{BC} &= a\overline{V}_{bc} \\ \overline{V}_{CA} &= a\overline{V}_{ca} \end{aligned} \tag{4.16}$$

The primary and secondary voltages are related by the winding turns ratio a, again just as in the wye–wye transformer. It is not possible to identify the line to ground voltages for the primary or secondary side voltage without further details on the power system connected to the respective winding. In the delta–delta transformer, neither winding is grounded. The winding could be grounded through a ground connection on the connected system, or it could be a part of a floating system. In either case, we will assume for the moment that the delta winding line to ground voltages are balanced, with equal magnitudes and separated in time by 120°. We label these voltages line to neutral voltages, although no neutral point is identified. The result for the primary side is that

$$\begin{aligned} \overline{V}_{AN} &= \frac{1}{\sqrt{3}} V_{AB}(1 \underline{/-30^\circ}) \\ \overline{V}_{BN} &= \frac{1}{\sqrt{3}} V_{BC}(1 \underline{/-30^\circ}) \\ \overline{V}_{CN} &= \frac{1}{\sqrt{3}} V_{CA}(1 \underline{/-30^\circ}) \end{aligned} \tag{4.17}$$

These line to neutral voltages can be used in the per phase equivalent circuit analysis for delta–delta transformers.

4.6 Per phase analysis of three phase transformers

Per phase analysis of balanced three phase systems was presented in chapter 1. In this chapter, we extend this concept to include balanced three phase transformers in the per phase analysis.

Per phase analysis assumes that the balanced system is a wye based system. In the case of delta connected sources or loads, both are transformed into the equivalent wye source or load.

The one line diagram of a balanced three phase system with a transformer is shown in figure 4.12.

One line diagrams are a convenient method of describing three phase systems. Each element is considered to be balanced and three phase.

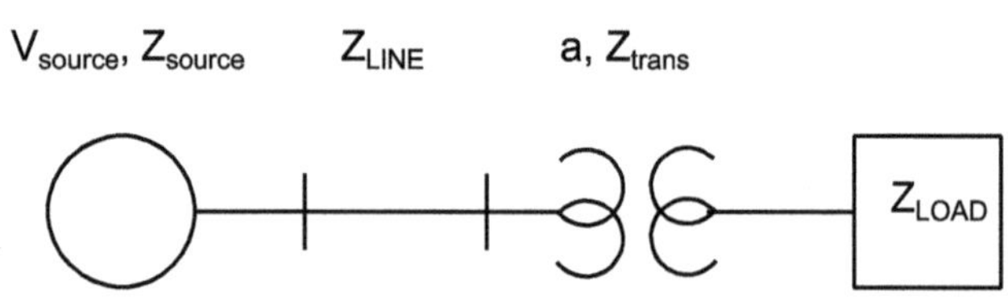

Figure 4.12. Generic one line diagram of a balanced three phase system with transformer.

It is important in three phase systems to be both concise and accurate in describing the three phase quantities involved. In specifying voltages, for example, it is not sufficient to say that the source has a magnitude of 4.16 kV. If V_{source} is a wye source, it would be stated as being 4.16 kV/2.4 kV wye or 4.16 kV/2.4 kV grid wye. On the other hand, a 4.16 kV delta source should be referred to as 4.16 kV delta. Notice that these full description of voltages is not always followed—in these cases, it is often important to check into the details of the source.

Transformers have primary and secondary windings, and are therefore still more complicated. The transformer voltages should be defined as shown in table 4.1, for the various types of winding. This table describes four types of transformer that step the voltage down from 4160 V line to line to 480 V line to line. The first set of numbers is the rated primary voltage along with an indication of winding type. Note that the transformer is designed to operate at voltages within ±5% of this value. The second set of voltages is the rated secondary voltage plus the indication of the type of secondary winding. The transformer turns ratio a is determined from the ratio of the primary to secondary winding voltage, as shown in the third column of the table. In the 4th column of the table, the so called 'bank' turns ratio is defined as the ratio of the line to line voltages of the transformer. The notation 'n' is used for the bank turns ratio in this book. The relationship between this bank ratio and the actual winding turns ratio is also shown in this column. The bank ratio is widely used in industry, particularly among those who analyse only balanced three phase systems. In the final column, the phase shift between primary and secondary line quantities is shown. In the delta–wye and wye–delta transformers, the direction of the 30° phase shift is determined by the specific connection of the delta.

Per phase analysis of a balanced three phase system relies on the fact that the phase voltages and currents will have equal magnitudes, and will be separated in time by 120°. As a result, when the A phase current at a point is known, the B and C phase currents can be inferred directly from this value. Similarly, when the A phase

Table 4.1. Primary and secondary voltage descriptions for common two winding transformer types.

Type of winding	Voltage statement	Winding turns ratio a	Bank turns ratio n (ratio of line to line voltages)	Phase shift between primary and secondary line quantities
Wye–wye	4.16 kV/2.4 kV wye: 480 V/277 V wye	$a = \frac{2400\ \text{V}}{277\ \text{V}} = 8.67$	$n = \frac{4160\ \text{V}}{480\ \text{V}} = a = 8.667$	0°
Delta–wye	4.16 kV delta: 480/277 V wye	$a = \frac{4160\ \text{V}}{277\ \text{V}} = 15$	$n = \frac{4160\ \text{V}}{480\ \text{V}} = \frac{a}{\sqrt{3}} = 8.667$	±30°
Wye–delta	4160 V/2400 V wye: 480 V delta	$a = \frac{2400\ \text{V}}{480\ \text{V}} = 5$	$n = \frac{4160\ \text{V}}{480\ \text{V}} = \sqrt{3}a = 8.667$	±30°
Delta–delta	4160 V delta: 480 V delta	$a = \frac{4160\ \text{V}}{480\ \text{V}} = 8.667$	$n = \frac{4160\ \text{V}}{480\ \text{V}} = a = 8.667$	0°

line to neutral voltage at a point is known, the B and C phase line to neutral values can be determined, as can the line to line values at that point.

The per phase diagram assumes a wye connected system, and represents A phase of that system. The wye–wye transformer is the most conceptually straightforward option. The one line diagram for a wye–wye transformer of figure 4.12 is shown in figure 4.13. In the wye–wye transformer, each phase is independent of the others, so the A phase circuit is simply the A to ground voltage connected to the transformer primary through the source side impedances $\overline{Z}_{\text{source}}$ and $\overline{Z}_{\text{line}}$. The ideal transformer is next, showing a turns ratio of a:1. The transformer impedance is represented here on the secondary side. Then the transformer secondary winding is connected directly to the load impedance $\overline{Z}_{\text{load}}$. The transformer impedance could be equally well represented on the primary side, with the primary equivalent impedance used rather than the secondary impedance.

The circuit in figure 4.13 is a single loop circuit, even though it does not appear to be. This is because the primary and secondary currents are related by the turns ratio, a known constant. It is sometimes easiest to refer all quantities to either the primary or secondary side. The secondary side version of figure 4.13 is shown in figure 4.14. An alternate way to get this diagram is to take the Thevenin equivalent of the source, looking into the system from the load terminals.

In the delta–wye transformer and the wye–delta transformer, it is easiest to represent the transformer impedance in the per phase diagram on the wye side of the transformer. The turns ratio to use in the diagram is the bank turns ratio. The phase shift can be represented by a unit vector with a 30° shift. The resulting per phase

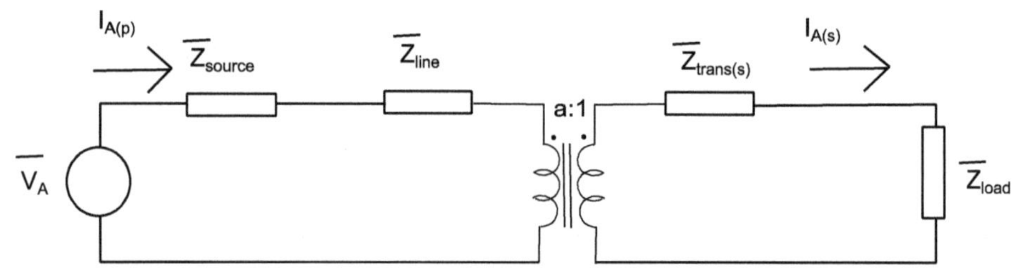

Figure 4.13. Per phase diagram for the circuit of figure 4.12 with a wye–wye transformer.

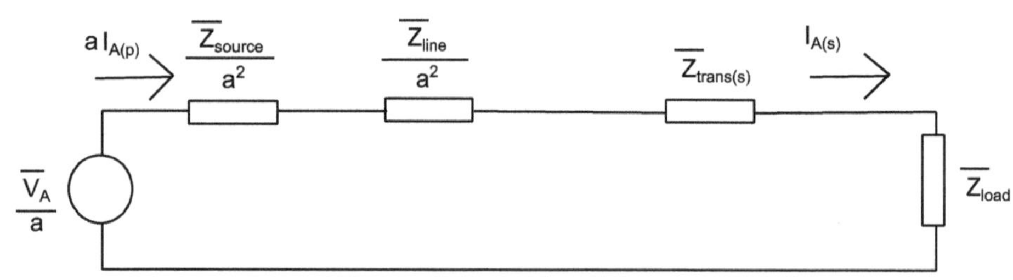

Figure 4.14. Per phase diagram of the system of figure 4.12, with voltage, current and impedances referred to the secondary side of the transformer.

diagram for the delta–wye transformer is shown in figure 4.15, for the case where the delta windings are connected A–B, B–C, and C–A as shown in figure 4.7.

This diagram is similar to figure 4.13, apart from the ratio of transformer turns. In this case with the delta–wye transformer, the per phase diagram relates primarily to secondary line currents, so the bank ratio n is used for the per phase diagram turns ratio. Figure 4.15 also shows the phase shift with the ratio being complex, with a $-30°$ phase shift.

In practice, this angle difference is often neglected in load and voltage studies on balanced systems. There are some instances, however, where the phase angle shift is very important.

In the delta–delta transformer, the transformer impedance does not see the line current on either winding. As a result, the transformer impedance must be adjusted in the per phase diagram.

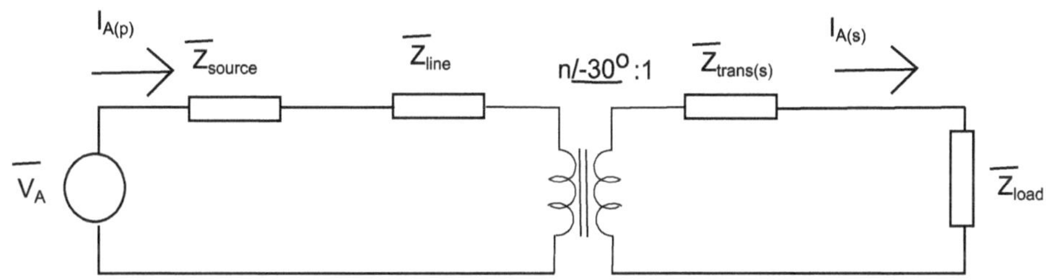

Figure 4.15. Per phase diagram for the system of figure 4.12 with a delta–wye transformer.

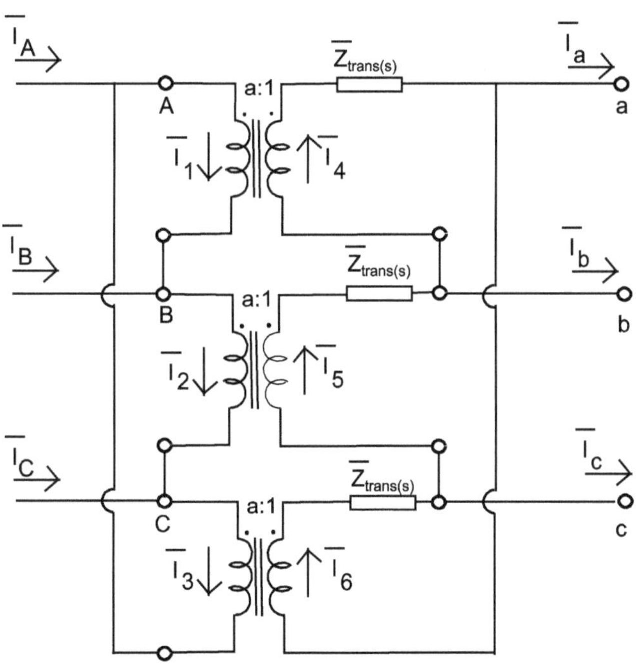

Figure 4.16. Delta–delta transformer showing transformer impedance on the secondary side.

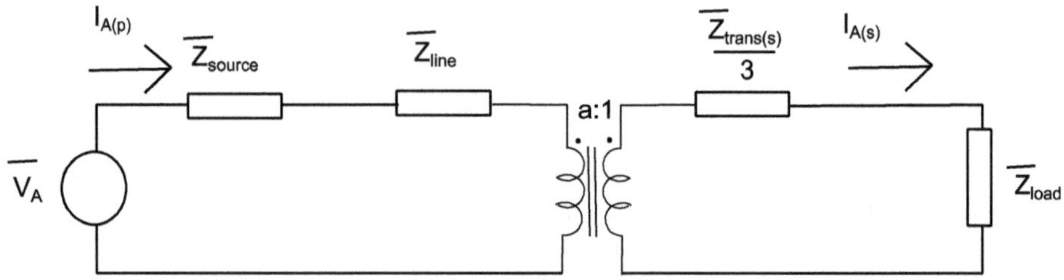

Figure 4.17. The per phase diagram for a delta–delta transformer.

In order to use the per phase equivalent circuit with this transformer connection, the transformer impedance must be converted to an equivalent impedance in the lines. Using superposition, the delta to wye impedance conversion of chapter 1 can be used to find that the equivalent secondary line impedance is

$$\frac{\overline{Z}_{\text{trans(s)}}}{3} \tag{4.18}$$

The resulting per phase equivalent circuit for the delta–delta transformer is shown in figure 4.17. Note also that the load impedance $\overline{Z}_{\text{load}}$ is the wye equivalent impedance of the load. In the delta–delta case, it is important to analyse the impedances closely when using the per phase equivalent circuit.

Questions

1. Do a web search, and find the meaning of the acronym FOFA as it relates to power transformers.
2. Search the term 'three phase power transformer vector group.' Describe the meaning of this term.

Problems

1. Three single phase transformers each have a turns ratio of three. The low voltage winding has a voltage rating of 2400 V. The secondary current rating is 150 A.
   a. The transformers are connected wye–wye. Find the line to line voltages of the high and low voltage sides when the transformers have balanced three phase voltages of 2400 V on the three secondary windings.
   b. Determine the primary line currents when the transformer low side is delivering balanced rated current to a load.
2. Repeat problem 1 when the high voltage side is connected in delta, and the low side windings are wye connected. The transformer connection is the one shown in figure 4.7.
3. The single phase transformers of problems 1 and 2 have a series impedance of $0.25 + j1.5$ ohms referred to the low side winding. The magnetizing branch of the transformer is negligible. A balanced three phase wye connected load of $\overline{Z}_{\text{wye}} = 50\ \Omega$ is connected to the low side of the transformer.

a. Draw the per phase equivalent circuit of the transformer.
b. In this case, $\overline{Z}_{source} = \overline{Z}_{line} = 0$. The source side line to neutral voltage magnitude is 7300 V. Assume that the A phase high side voltage is at an angle of 0°. Find the A phase line to neutral voltage on the low voltage winding.
c. What are the real and reactive powers entering the high side of the three phase transformer bank?

4. Repeat problem 3 with $\overline{Z}_{source} = \overline{Z}_{line} = j8\ \Omega$.

IOP Publishing

Electromechanical Machinery Theory and Performance

Thomas Ortmeyer

Chapter 5

Rotating AC machine basics

In this chapter, the basic structure of AC machines is investigated. In particular, the air gap flux patterns and development of torque are discussed.

5.1 The two pole one phase machine

Figure 5.1 shows a three dimensional view of a simple synchronous machine rotor. The rotor has active length ℓ_c and radius r_{gap}. The coil on the rotor has number of turns N_f and carries current i_f. While it is shown as occupying a single pair of slots, in practice this coil will be distributed in several slots around the rotor, in a manner that optimizes the machine design.

A conceptual cross sectional diagram of a simple synchronous machine is shown in figure 5.2. It has a single coil on the rotor, and one single coil on the stator. The reference direction for current in the coil is that current comes out of the dotted coil side (tip of the arrow) and current goes into the crossed coil side (back end of the arrow). The rotor coil carries direct current, and establishes a magnetic field. (In some machines, this field is set up by permanent magnets.) This field sets up a magnetic flux that crosses the air gap and links the stator coil, which we will refer to as the A phase coil. Figure 5.2 shows the position of the A phase coil in one pair of slots on the stator. Positive current in this coil will create a magnetic field with a peak flux density perpendicular to the plane of the coil. This is labeled as the A phase magnetic axis in the diagram. The field coil on the rotor is also shown as occupying a single set of slots on the rotor. The field magnetic axis is perpendicular to the plane of this coil, and is shown in the figure. Note that in practice, this A phase coil will also be distributed around the interior stator perimeter, just as the field coil is distributed around the rotor perimeter.

As shown in figure 5.2, we define the position of the rotor as the angle θ between the field magnetic axis and the A phase magnetic axis. Note that as the A phase coil is fixed to the stator, so is its magnetic axis and it is therefore stationary. Similarly,

doi:10.1088/978-0-7503-1662-0ch5

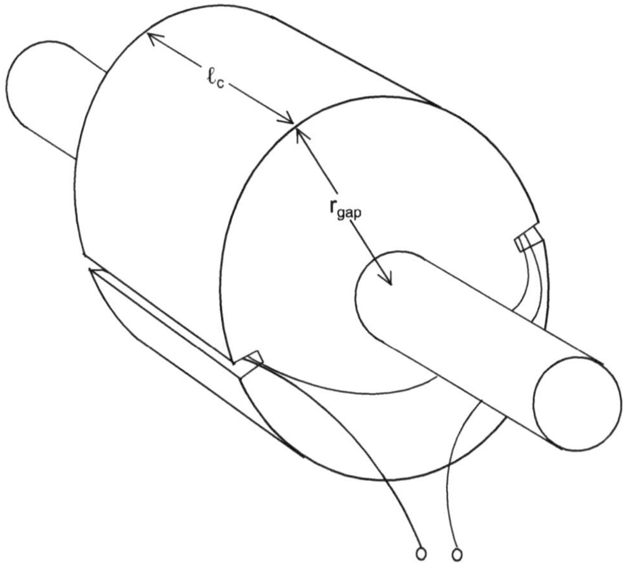

Figure 5.1. Three dimensional view of machine rotor with a single coil.

the field coil is fixed to the rotor, as is the field magnetic axis, and both move with the rotor as it rotates on its shaft.

It is necessary to write an equation for flux density as a function of the angular position in the air gap. Figure 5.2 shows an arbitrary position in the air gap, which is at an angle γ from the A phase magnetic axis or an angle ρ as measured from the field magnetic axis. Therefore,

$$\gamma = \theta + \rho \tag{5.1}$$

Current in the field winding sets up a flux density wave in the air gap. Ideally this flux wave is sinusoidally distributed around the air gap. This sinusoidal field flux distribution is created through careful design of the air gap and/or the distribution of the field winding. In this case, the peak of the flux density wave created by the field current occurs on the field magnetic axis, and is positive when the field current is positive. This flux density in the air gap can be written as

$$B_{\mathrm{F}}(\rho) = B_{\mathrm{Fmax}} \cos(\rho) \tag{5.2}$$

In equation (5.2), the flux density is measured with respect to the field magnetic axis, which rotates as the rotor turns. The positive peak flux density is always at $\rho = 0°$. Positive values indicate flux that is going from the rotor to the stator. At $\rho = 180°$, the flux density is at its negative peak, which indicates flux traveling from stator to rotor. As lines of flux must be closed, the total flux passing from rotor to stator (in angles $-90° < \rho < 90°$) must equal the total flux passing from stator to rotor (when $90° < \rho < 270°$).

When flux density is measured with respect to the stationary A phase magnetic axis, it is (from equation (5.1))

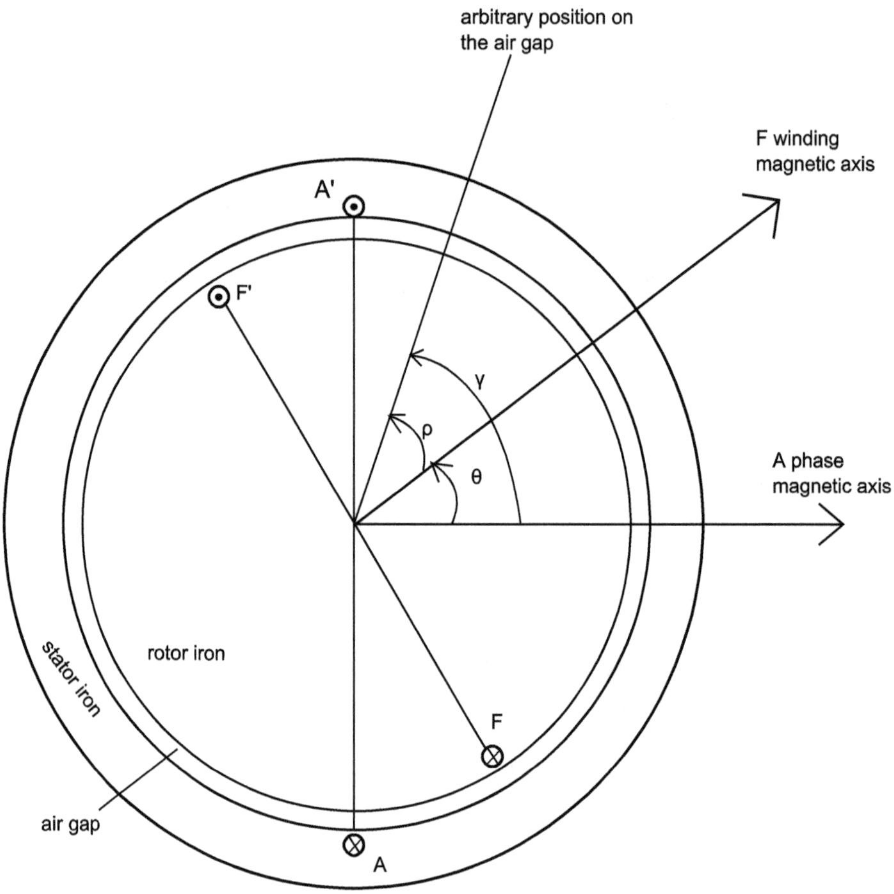

Figure 5.2. Simple machine cross section showing field winding F on the rotor and A phase winding A on the stator.

$$B_F(\gamma) = B_{\text{Fmax}}\cos(\gamma - \theta) \tag{5.3}$$

The angle θ is the angle between the rotor and the stator.

Example 5.1. For figure 5.3, a current flows in the field winding so that $B_{\text{Fmax}} = 1$. The rotor is spinning at a constant speed of 10 radians/second, with a rotor position of $\theta = 0°$ at time $t = 0$ s. What is the machine flux density created by this field current, with reference to the stator magnetic axis?

Solution. From equation (5.2), the flux density measured with reference to the rotor magnetic axis is $B_F(\rho) = \cos(\rho)$. With a rotor speed of $\omega_r = 10\frac{\text{rad}}{\text{sec}}$, the rotor position is

$$\int_o^\theta d\theta = \int_0^t \omega_r dt$$

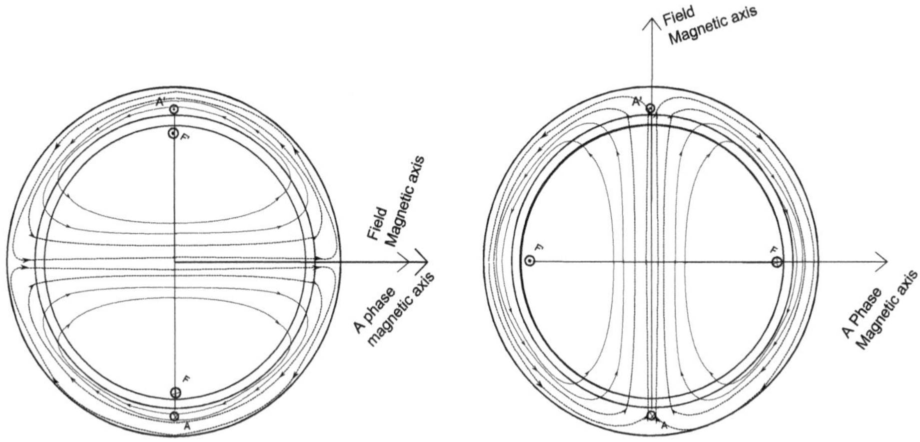

(a) $\theta = 0°$, magnetic axes aligned. (b) $\theta = 0°$, magnetic axes perpendicular

Figure 5.3. Field flux patterns for various rotor positions.

Giving

$$\theta = 10t$$

So the field flux density measured with respect to the stator will be

$$B_F(\gamma) = \cos(\gamma - 10t)$$

When viewed from a fixed position on the stator, say $\gamma = 0$, the flux density will be a sinusoidal function of time. Furthermore, the frequency of this time variation will be the same as the speed of the rotor.

It is important to determine the total field flux that will link the A phase winding. When the rotor position is $\theta = 0°$, the field magnetic axis and the A phase magnetic axis will be aligned. Field flux will be positive (going from rotor to stator) in the right hand side of the stator. It will circulate through the stator back iron and to the left hand side of the stator, and then will flow back across the air gap from stator to rotor. The total amount of flux crossing from rotor to stator on the right must equal the amount of flux passing from stator to rotor on the left. Any flux passing from rotor to stator on the right hand side of the motor will link the A phase winding. At $\theta = 0°$, the field flux will be at a maximum on the A phase magnetic axis, and the field flux will be positive throughout the right half of the motor. This will be the rotor position where the flux linking the A phase winding will be a maximum. This case is shown in figure 5.3(a). When θ increases, however, the field flux pattern will rotate, and the A phase winding will see less flux. At $\theta = 45°$, the peak field flux density will be at $\gamma = 45°$, and the flux density will equal zero at angles of $\gamma = -45°$ and $\gamma = 135°$. In both cases, the field flux linking the A phase winding will result from the flux passing between $\gamma = -90°$ and $\gamma = 90°$, as the A phase coil has not moved. The field flux linking the A phase coil can be determined by integrating the flux density over the area to the right of the plane of the A phase coil,

$$\Phi_{Af}(\theta) = \ell_c r_{gap} \int_{-\pi/2}^{\pi/2} B(\gamma, \theta) d\gamma \tag{5.4}$$

Here, r_{gap} is the effective radius of the machine air gap, which is slightly larger than the radius of the rotor itself. ℓ_c is the cylindrical length of the rotor. The result of this integration is

$$\Phi_{Af}(\theta) = 2\ell_c r_{gap} B_{max} \cos(\theta) \tag{5.5}$$

It is important to understand the relationship between Φ_{Af} and rotor position θ. It has been noted that Φ_{Af} is maximum when $\theta = 0°$. When $\theta = 90°$, the field flux will be oriented vertically (in parallel with the A coil), and no flux will pass through the plane of this coil. The result is that there will be no field flux linking the A phase winding when the rotor position $\theta = 90°$.

At $\theta = 180°$, the field winding will be in a similar position to $\theta = 0°$, but with the polarity reversed. As a result, the full field flux will link the A phase winding, but in the reverse (negative) direction. When the rotor rotates the field to $\theta = 270° = -90°$, the field flux linking the A winding will again be zero. As equation (5.5) states, this variation of flux linking the A phase coil will be sinusoidal.

Figure 5.3 shows the field flux pattern at angles of $\theta = 0°$ and $\theta = 90°$. With a little imagination, the flux patterns for the other 358° can be imagined.

There will be a voltage induced in coil A due to this changing flux linkage from the field. If coil A has N_s turns, this voltage will be

$$e_{Af} = N_s \frac{d\Phi_{Af}}{dt} \tag{5.6}$$

or

$$e_{Af} = N_s 2\ell_c r_{gap} B_{max} \frac{d(\cos\theta)}{dt} = -N_s 2\ell_c r_{gap} B_{max} \frac{d\theta}{dt} \sin\theta \tag{5.7}$$

If the rotor is operating at constant speed ω_r, then $\omega_r = \frac{d\theta}{dt}$. It is convenient to select the time scale so that $\theta = \omega_r t - 90°$. Then

$$e_{Af} = N_s 2\ell_c r_{gap} B_{max} \omega_r \sin\omega_r t \tag{5.8}$$

Under these conditions (constant speed, sinusoidal flux distribution), the induced voltage in the winding will be sinusoidal with the electrical frequency in electrical radians per second equal to the mechanical speed in mechanical radians per seconds, $\omega_e = 2\pi f_e = \omega_r$. The magnitude of this voltage will be proportional to both the magnitude of the field flux linking the coil and the rotor speed.

The term 'synchronous' machine comes from this linkage of electrical frequency and mechanical speed. This relationship can also be expressed as

$$f_e = \frac{n_r}{60} \tag{5.9}$$

where f_e is the electrical frequency in units of hertz and n_r is the mechanical shaft speed in units of rpm (revolutions per minute).

5.2 Higher pole machines

Figure 5.4 shows a cross section diagram of a three phase two pole machine. For convenience, the machine is shown as having salient rotor poles, rather than a round rotor. The air gap is represented as a circle, with the stator outside the air gap and holding the three phase windings. The field (F) winding will carry a direct current, which will create a field flux that flows out of the rotor in the direction of the field magnetic axis, through the stator iron, and then back into the opposite end of the rotor to complete the flux path. In one rotor revolution, the A phase coil will see one full cycle of field flux linkage—the A phase coil will see each rotor pole face one time for one mechanical revolution. One of these pole faces will create positive flux linkage in the A coil, the other will create negative flux linkage in this coil. The field flux linking the A coil on this machine is therefore

$$\lambda_{Af} = \lambda_{MAX} \cos\theta = N_s \Phi_{AfMAX} \cos\theta \tag{5.10}$$

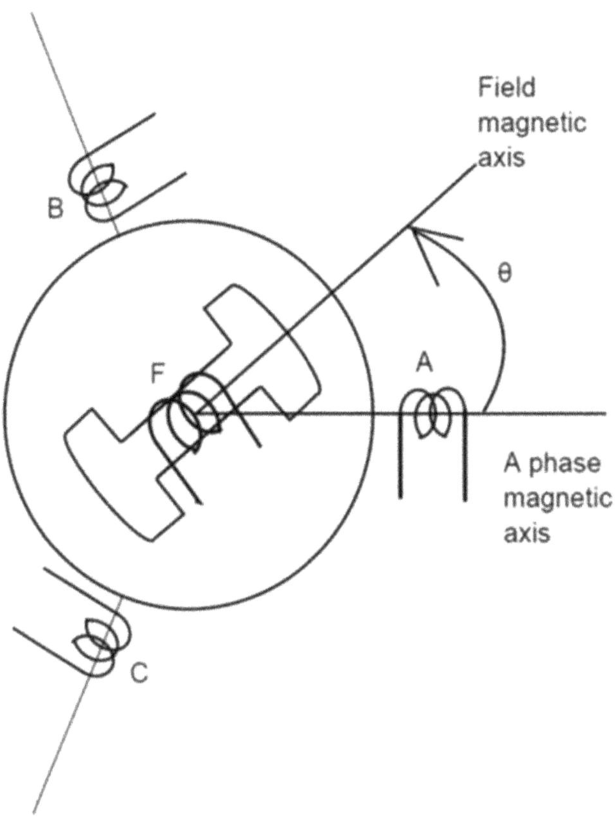

Figure 5.4. Cross section of a two pole three phase machine with field winding F on the rotor.

where Φ_{AfMAX} is the maximum value of the flux density from equation (5.5). Keep in mind the flux density B_{max} varies with field current.

Many machines have more than two poles—but these poles must come in pairs. Figure 5.5 shows a four pole machine, where the flux distribution has two full cycles around one revolution on the air gap. The salient pole rotor has four pole faces. On two of these poles, the field flux goes out of the rotor, and on the other two poles, the field flux re-enters the rotor body. The flux direction is shown by the arrows on the rotor poles.

On the stator, there are similarly two coils for each phase. All six of these coils will be identical, apart from their physical position along the air gap. The two coils in a given phase, say A and A′, having the same number of turns, can be connected in either series or parallel.

As figure 5.5 shows, four field poles will pass a given stator phase coil for each full revolution of the rotor.

For this reason on the four pole machine, the relationship between rotor position θ_r and A coil flux linkage is

$$\lambda_{Af} = \lambda_{MAX} \cos 2\theta_r \tag{5.11}$$

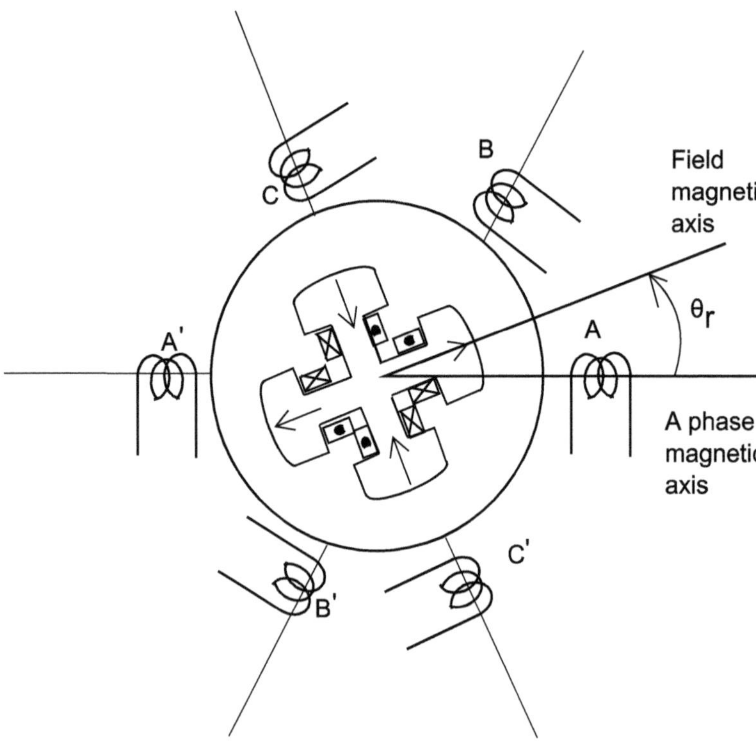

Figure 5.5. Four pole salient pole synchronous machine. Arrows on rotor field poles show the flux direction.

In general, for a machine with multiple poles, the field flux linking the A phase coil will be

$$\lambda_{\mathrm{Af}} = \lambda_{\mathrm{MAX}} \cos \frac{\mathrm{poles}}{2} \theta_{\mathrm{r}} \tag{5.12}$$

where 'poles' is the number of poles of the machine. The number of poles in this type of machine must be an even integer.

(Note: many texts use the symbol 'P' for poles—this can cause confusion when 'P' is used as the symbol for power. Hence the notation 'poles' is used here.)

It is convenient to define an electrical angle θ_{e}, that corresponds to the point in the electrical cycle rather than in the mechanical position of the motor. The subscript 'e' is used to denote electrical quantities. The mechanical angle is now called θ_{r}, with the subscript 'r' indicating that it is the physical position of the rotor. This electrical angle is

$$\theta_{\mathrm{e}} = \frac{\mathrm{poles}}{2} \theta_{\mathrm{r}} \tag{5.13}$$

The flux linking the A coil due to the field flux is then

$$\lambda_{\mathrm{Af}} = \lambda_{\mathrm{MAX}} \cos \theta_{\mathrm{e}} \tag{5.14}$$

The time derivative of the electrical angle θ_{e} will give the electrical frequency of the machine phase coils,

$$\omega_{\mathrm{e}} = \frac{d\theta_{\mathrm{e}}}{dt} = \frac{d}{dt}\left(\frac{\mathrm{poles}}{2}\theta_{\mathrm{r}}\right) = \frac{\mathrm{poles}}{2}\omega_{\mathrm{r}} \tag{5.15}$$

Here, ω_{r} is the rotor speed in radians per second. The electrical frequency is therefore directly related to the mechanical speed in rpm by the formula

$$f_{\mathrm{e}} = \frac{\mathrm{poles}}{2}\frac{n_{\mathrm{r}}}{60} \tag{5.16}$$

This is the general form of equation (5.9), correct for any number of pairs of poles.

Example 5.2. A six pole synchronous machine is operated from a 60 Hz power source. What is its mechanical speed?

Solution. From equation (5.6), the mechanical speed in rpm is

$$n_{\mathrm{r}} = \frac{120}{\mathrm{poles}} f_{\mathrm{e}} = 1200 \text{ rpm}$$

Example 5.3. A four pole synchronous machine is driven by a mechanical shaft that rotates at speeds between 1000 rpm and 2500 rpm. What will the output frequency of the machine be?

Solution. As the shaft speed varies, the electrical frequency will vary. From equation (5.16), the minimum frequency will be

$$f_e = \frac{4}{2}\frac{1000}{60} = 33.33 \text{ Hz}$$

Also, the maximum frequency will be

$$f_e = \frac{4}{2}\frac{2500}{60} = 166.7 \text{ Hz}$$

So the output frequency of this motor will be

$$33.33 \text{ Hz} \le f_e \le 166.7 \text{ Hz}$$

5.3 Three phase machines

Most synchronous machines have three stator phases. Figure 5.4 shows a two pole three phase machine, and figure 5.5 shows a four pole three phase machine. In general, the B and C phase coils are identical to the A phase coil, but are located at 120° and 240° electrical degrees from the A phase coil. In figure 5.4, the peak flux linkage for the B phase occurs 120° mechanically after the A phase peak, while in figure 5.5, the peak flux linkage for the B phase occurs 60° mechanically after the A phase peak. For the C phase coil, the peak flux linkage is 120° electrically and 60° mechanically after the B phase coil for the four pole machine. In two pole machines, the electrical angle θ_e equals the mechanical angle θ_r. In machines with higher numbers of pole pairs, such as the four pole machine shown in figure 5.5, equation (5.13) shows the relationship between the electrical and mechanical angles in the machine, $\theta_e = \frac{\text{poles}}{2}\theta_r$.

In general, the voltages induced in the A, B and C phase coils by the field flux are given in equation (5.17).

$$\begin{aligned}
e_{Af} &= \frac{d\lambda_{Af}}{dt} = \frac{d}{dt}(\lambda_{MAX}\cos\theta_e) = -\frac{d\theta_e}{dt}\lambda_{MAX}\sin\theta_e \\
e_{Bf} &= \frac{d\lambda_{Bf}}{dt} = \frac{d}{dt}\left(\lambda_{MAX}\cos\left(\theta_e - {}^{2\pi}\!/_3\right)\right) = -\frac{d\theta_e}{dt}\lambda_{MAX}\left(\theta_e - {}^{2\pi}\!/_3\right) \\
e_{Cf} &= \frac{d\lambda_{Cf}}{dt} = \frac{d}{dt}\left(\lambda_{MAX}\cos\left(\theta_e - {}^{4\pi}\!/_3\right)\right) = -\frac{d\theta_e}{dt}\lambda_{MAX}\left(\theta_e - {}^{4\pi}\!/_3\right)
\end{aligned} \tag{5.17}$$

In these equations, the voltages are written as a function of the electrical angle θ_e. It is worth noting that the induced voltages are out of phase with the field flux vector by 90°.

The angle θ_r is directly related to the physical position of the rotor, as shown in figure 5.5. When the rotor is turning at constant speed $\omega_r = \frac{d\theta_r}{dt}$ radians per second, the electrical angle θ_e becomes

$$\theta_e = \omega_e t + \delta_e = \frac{\text{poles}}{2}(\omega_r t + \delta_r) \quad (5.18)$$

In this equation, it can be seen that the electrical frequency of ω_e radians per second and the mechanical speed of ω_r radians per second are related by a factor of $\frac{\text{poles}}{2}$. The mechanical angle δ_r is the physical position of the rotor at time $t = 0$, and the electrical angle of the phasor voltage δ_e is related to this position scaled by the pole pair factor.

These machines are called synchronous machines because of this exact relationship between rotor position/speed and electrical angle/frequency.

Example 5.4. A three phase four pole synchronous machine is rotating at 1800 rpm. It has a flux linkage created by a rotor field winding of $\lambda_{MAX} = 0.12$ weber − turns. Find the instantaneous voltages on the three stator windings.

Solution. The stator electrical frequency will be

$$f_e = \frac{4}{2}\frac{1800\text{ rpm}}{60\text{ s/min}} = 60\text{ Hz}$$

From equation (5.18),

$$\theta_e = (2\pi 60)t + \delta_e = 377t + \delta_e$$

From equation (5.17), the A phase voltage will then be

$$\begin{aligned} e_{Af} &= -377(0.382)\sin(377t + \delta_e) \\ &= -144 \cdot \sin(377t + \delta_e) \\ &= 144 \cdot \cos(377t + \delta_e + \pi/2) \\ &= \sqrt{2} \cdot 102 \cdot \cos(377t + \delta_e + \pi/2) \end{aligned}$$

The A phase voltage is shown in several equivalent forms. As voltages are often expressed in terms of the cosine function rather than the sine function, these versions are shown. The final expression emphasizes that the RMS value of the generated voltage is 102 V. The phase angle in these expressions is set by the arbitrary choice of the mechanical angle reference in figure 5.3. As the angle measurements in any electrical circuit are relative, other angle references will often be chosen.

With this angle reference, the B and C phase voltages will the

$$\begin{aligned} e_{Bf} &= \sqrt{2} \cdot 102 \cdot \cos(377t + \delta_e - \pi/6) \\ e_{Cf} &= \sqrt{2} \cdot 102 \cdot \cos(377t + \delta_e + 5\pi/6) \end{aligned}$$

As expected, these voltages have the same magnitude, and are separated in time by 120°.

As in previous chapters, the phasor values of the voltages are expressed in terms of the RMS magnitude of the voltage, with the angle in degrees

$$\overline{E}_{Af} = 102\text{ V}\,\underline{/\delta_e + 90^\circ}$$
$$\overline{E}_{Bf} = 102\text{ V}\,\underline{/\delta_e - 30^\circ}$$
$$\overline{E}_{Cf} = 102\text{ V}\,\underline{/\delta_e - 150^\circ}$$

5.4 Stator current and flux

In a synchronous generator, the stator coils will be connected to loads, and currents will flow. The phase currents will in turn create magnetic flux distributed around the air gap. The stator phase windings are designed so that the distribution of this flux around the air gap will be sinusoidal. The stator A phase flux will be a function of the air gap position γ_e, where $\gamma_e = \frac{\text{poles}}{2}\gamma_r$ is the arbitrary position on the airgap in electrical radians, as defined in figure 5.3. The resulting flux linking the A phase coil due to the current in the A phase coil is

$$\Phi_{As} = K_s i_A \cos \gamma_e \tag{5.19}$$

As this coil is stationary, the flux will always have a peak at $\gamma_e = 0$, on the A phase magnetic axis. It will vary with the level of the instantaneous current i_A. The subscript 's' indicates flux created by stator current. The constant K_s is a function of several parameters, including the rotor length, rotor radius, air gap length, winding factor, etc, and can be derived in a similar method as was done for the field current in equation (5.5).

The flux created by the B and C coil currents will be

$$\begin{aligned}\Phi_{Bs} &= K_s i_B \cos\left(\gamma_e - {}^{2\pi}/_3\right)\\ \Phi_{Cs} &= K_s i_C \cos\left(\gamma_e - {}^{4\pi}/_3\right)\end{aligned} \tag{5.20}$$

Similar to the A phase, the B phase flux will always have its maximum value on the B phase magnetic axis, $\gamma_e = {}^{2\pi}/_3$, and the C phase flux will always have its maximum on the C phase magnetic axis, $\gamma_e = {}^{4\pi}/_3$ or 240°.

In the steady state when the loads are balanced, the currents will also be balanced,

$$\begin{aligned}i_A &= \sqrt{2}I_s \cos(\omega_e t + \phi_e)\\ i_B &= \sqrt{2}I_s \cos\left(\omega_e t + \phi_e - {}^{2\pi}/_3\right)\\ i_C &= \sqrt{2}I_s \cos\left(\omega_e t + \phi_e - {}^{4\pi}/_3\right)\end{aligned} \tag{5.21}$$

By combining equations (5.19) through (5.21), the air gap flux created by these balanced stator phase currents will be

$$\begin{aligned}
\Phi_{As} &= \sqrt{2}K_s I_s \cos(\omega_e t + \phi_e)\cos\gamma_e \\
\Phi_{Bs} &= \sqrt{2}K_s I_s \cos\left(\omega_e t + \phi_e - {}^{2\pi}\!/_{3}\right)\cos\left(\gamma_e - {}^{2\pi}\!/_{3}\right) \\
\Phi_{Cs} &= \sqrt{2}K_s I_s \cos\left(\omega_e t + \phi_e - {}^{4\pi}\!/_{3}\right)\cos\left(\gamma_e - {}^{4\pi}\!/_{3}\right)
\end{aligned} \tag{5.22}$$

The standing flux waves of each coil: Individually, each phase will create a standing flux wave around the machine air gap with peak amplitude in space on the magnetic axis of the respective phase (at $\gamma_e = 0$). The amplitude of this standing wave will vary with time, as the phase currents vary according to equation (5.21).

The rotating flux wave of the combined stator currents: The total flux induced by the three phase currents on the stator will add algebraically at points around the air gap (assuming that core saturation is negligible). The total stator flux will then be

$$\Phi_S = \Phi_{As} + \Phi_{Bs} + \Phi_{Cs} \tag{5.23}$$

With balanced currents in each phase (as in equation (5.21)), the stator flux wave can be found from equations (5.22) and (5.23). The result is

$$\Phi_S = \frac{3K}{\sqrt{2}} I_s \cos(\omega_e t + \phi_e - \gamma_e) \tag{5.24}$$

Equation (5.23) is the equation of a rotating wave. Φ_S has a constant magnitude of $\frac{3K_S}{\sqrt{2}} I_s$ and having peak flux at the rotor position $\gamma_e = \omega_e t + \phi_e$. This rotating flux wave will be rotating at the same speed ($\omega_e = \frac{\text{poles}}{2}\omega_r$) as the rotor. This is a result of the stator phase currents having the same frequency as the voltage induced by the rotor field flux.

The significance of this point cannot be overstated—injecting three phase currents that are separated by 120° in time into three phase windings that are separated 120° in space creates a rotating flux wave of constant magnitude and rotational speed equal to electrical frequency. In this case, the machine will have a constant power flowing at its electrical terminals, and will develop a constant torque on the shaft.

5.5 Synchronous generator per phase equivalent circuit

The A phase winding is linked by both the field and the stator flux waves. Both flux waves have constant magnitude and are rotating at the same speed, ω_e in electrical radians per second.

The field flux induces the voltage e_{Af} in winding A, as described in equation (5.17). The rotating flux wave created by the three stator currents will induce a voltage in the A phase stator winding proportional to the time derivative of the stator flux Φ_s linking the A phase coil. This voltage is often termed the armature reaction voltage, and is proportional to the magnitude of the stator current I_s and the frequency of the stator currents ω_e. The proportionality constant is the synchronous inductance L_s. The time derivative will produce a voltage that lags the current by 90° when the currents are sinusoidal. As a result, the A phase winding will have a sinusoidal voltage induced by the field flux, and will be modified by an additional voltage induced by flux created by the stator current. The resulting per phase equivalent circuit of the stator is shown in figure 5.6.

In this figure, the phasor voltage $\overline{E}_{Af}$ is the machine internal voltage, which is induced by the field flux. The synchronous inductance L_s represents the effect of the armature reaction as well as the stator phase leakage inductance. When the machine is operating at frequency ω_e, the synchronous reactance of the machine will be

$$X_s = \omega_e L_s$$

The resistance R_s represents the stator coil resistance and any stray loss effect. The terminal voltage $\overline{V}_t$ is a function of all three of these parameters plus the stator current $\overline{I}_S$. As noted in the figure, the reference direction for real and reactive power is the same direction as the current reference.

The magnitude of the armature phase internal voltage E_{Af} is determined by the level of the field flux, the speed of machine rotation, and the coupling between field and stator winding. In practice, there are two types of synchronous machine field:

- permanent magnet field,
- wound field.

For permanent magnet machines, the field flux is constant, and the induced voltage is

$$E_{Af} = K_{pm}\omega_e \tag{5.25}$$

In the case of wound field machines, the level of flux will increase as field current is increased. As the flux increases, however, the machine core will start to saturate. Typically, these machines are operated with some level of saturation. The saturation is a function of the total machine flux, however, not just the field flux.

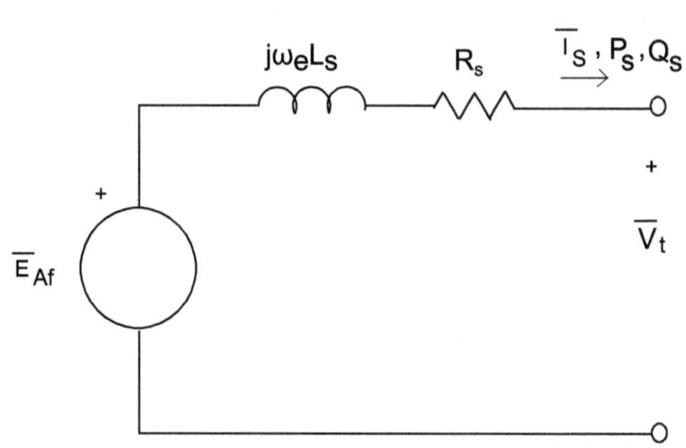

Figure 5.6. Per phase equivalent circuit of a three phase synchronous generator.

In practice, the open circuit excitation curve of a synchronous generator is often measured by the manufacturer during machine commissioning. A typical curve is shown in figure 5.7.

In this curve, the machine terminal voltage is recorded as the field current is varied during open circuit armature conditions at rated machine speed. The results are reported as % of rated voltage—this percentage can be applied to either line to line or line to neutral voltages. The field current is plotted as percent of nominal, with nominal field current providing rated voltage at rated speed.

Two additional lines have been added to the plot—the first is the air gap line, which is valid up to levels where core saturation has an impact. The second added (dashed) line is the linear approximation of the machine that is valid when the machine terminal voltage is near the rated value. Keep in mind that saturation of a round rotor machine is a function of the total air gap flux rather than the field flux.

The induced phase voltage E_{Af} is proportional to the machine rotational speed ω_e, the field current I_F, and the machine reactance. In situations where the air gap flux is

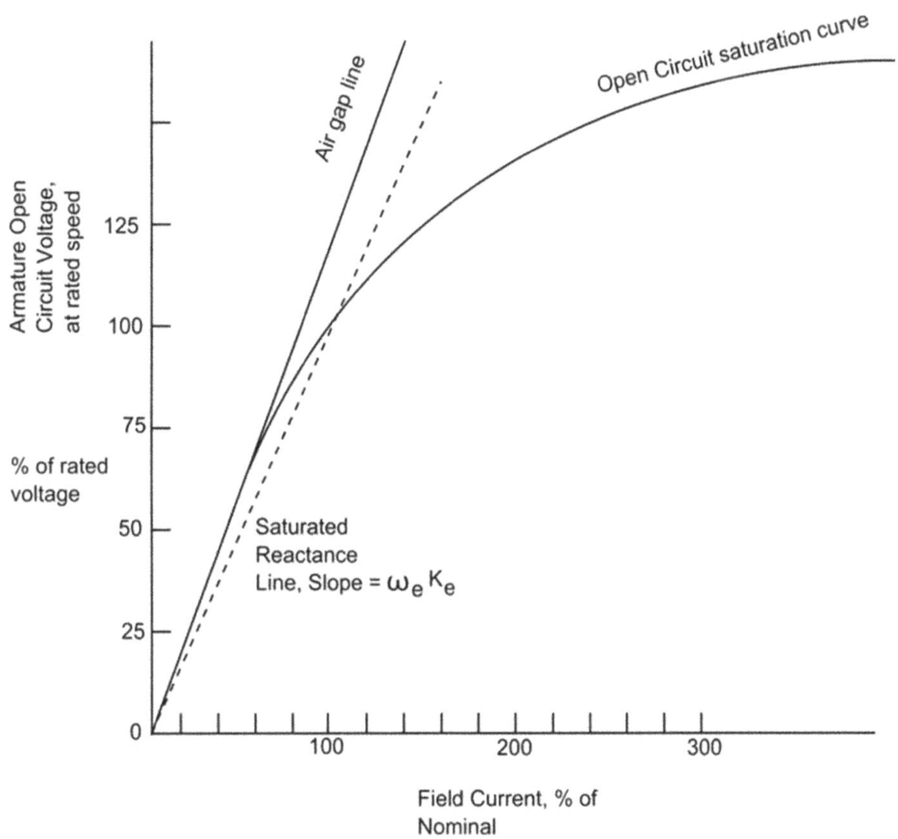

Figure 5.7. Synchronous machine open circuit excitation curve.

close to its rated value, it is sufficiently accurate to use the saturated reactance value, so that

$$E_{Af} = \omega_e K_e I_F \tag{5.26}$$

Example 5.5. A three phase four pole synchronous generator has ratings 150 MVA, 60 Hz, 13.2 kV/7.62 kV wye. When the generator is operating at rated speed and frequency, a field current of 140 A is required to develop rated open circuit voltage. The generator synchronous reactance X_s = 1.12 ohms. The resistance of the generator phases can be neglected.

(a) What is the rated generator speed?

In radians per second, the machine speed is $\omega_r = \omega_e/(\frac{\text{poles}}{2}) = \frac{2\pi 60}{4/2} = 188.5 \frac{\text{radians}}{\text{sec}}$

In rpm, from equation (5.16), $n_m = \frac{60 f_e}{\text{poles}/2} = \frac{3600}{2} = 1800$ rpm

(b) The generator is connected to an infinite bus which has a voltage of 13.2 kV/7.62 kV wye. Draw the per phase equivalent circuit of the generator.

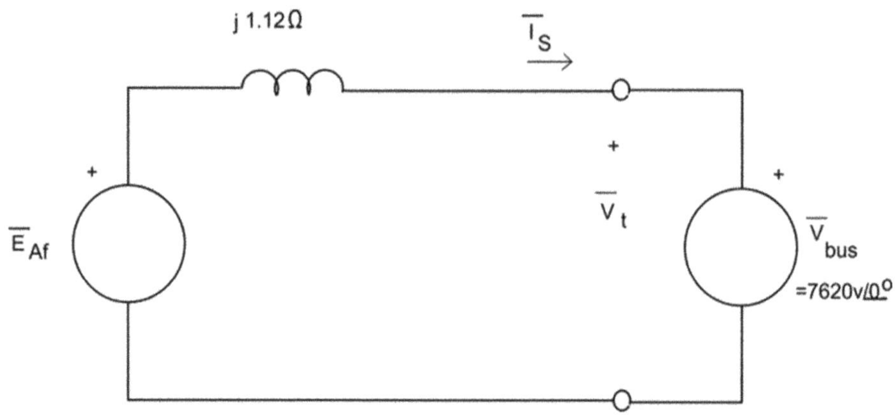

(c) The generator is supplying 80% of rated load at unity power factor. Calculate the magnitude and angle of the generator phase current.

$$I_{\text{rated}} = \frac{S_{3\phi\text{rated}}}{3V_{L-N\text{rated}}} = \frac{0.80 \cdot 150 \times 10^6 \text{ VA}}{3 \cdot 7620 \text{ V}} = 5249 \text{ A}$$

Power factor angle: $\theta_s = \delta_s - \phi_s = \cos^{-1} 1.0 = 0°$. Since the voltage angle δ_s was chosen to be 0°, $\phi_s = 0° - 0° = 0°$. The phase current phasor is therefore

$$\bar{I}_s = 5249 \text{ A} \,\underline{/0°}$$

(d) Find the internal line-neutral voltage $\overline{E}_{\mathrm{Af}}$:

$$\overline{E}_{\mathrm{Af}} = \overline{V}_{\mathrm{bus}} + jX_s\overline{I}_s = 7620\,\underline{/0^\circ} + j1.12(5249\ \mathrm{A}\,\underline{/0^\circ})$$
$$\overline{E}_{\mathrm{Af}} = 7620 + j5879 = 9624\ \mathrm{V}\,\underline{/37.8^\circ}$$

(e) Find the generator field current.

With the generator open circuited and operating at rated speed and rated voltage, the field current is given to be 140 A. Working with the line to neutral voltage, with $E_{\mathrm{Af}} = V_t$ at no load,

$$\omega_e K_e = \frac{E_{\mathrm{Af}}}{I_f} = \frac{7620\ \mathrm{V}}{140\ \mathrm{A}} = 54.43$$

With the loading of this example, E_{Af} will have the value found in (d), and

$$I_f = \frac{E_{\mathrm{Af}}}{\omega_e K_e} = \frac{9624\ \mathrm{V}}{54.43} = 177\ \mathrm{A}$$

Note that this field current is DC.

Example 5.6. Repeat example 5.5 with the same VA output of the generator, but a power factor of 0.85 lagging.

In this case, the angle between voltage and current is $\theta_s = \delta_s - \phi_s = \cos^{-1} 0.85 = 31.8^\circ$. As the power factor is lagging, θ_s will be positive. θ_s would be negative if the power factor is leading.

Since $\delta_s = 0^\circ$, $\phi_s = -31.8^\circ$. Then

$$\overline{E}_{\mathrm{Af}} = \overline{V}_{\mathrm{bus}} + jX_s\overline{I}_s = 7620\,\underline{/0^\circ} + j1.12(5249\ \mathrm{A}\,\underline{/-31.8^\circ})$$
$$\overline{E}_{\mathrm{Af}} = 7620 + j5879 = 9624\ \mathrm{V}\,\underline{/37.8^\circ}$$
$$\overline{E}_{\mathrm{Af}} = 10717 + j4998 = 11824\ \mathrm{V}\,\underline{/25.0^\circ}$$

It is interesting to compare this result with that of example 5.5, and note that the same current with a lagging power factor causes a significant rise in internal voltage magnitude E_{Af}, which requires a corresponding increase in field current,

$$I_f = \frac{11824\ \mathrm{V}}{54.43} = 217\ \mathrm{A}$$

Generator convention and motor convention

In figure 5.6, the current direction is chosen to be out of the winding. The three phase power flowing out of the machine is then

$$P_{3\phi} = 3V_tI_S\cos(\delta_t - \phi_s) \tag{5.27}$$

Here, the generator terminal voltage is $\overline{V}_t = V_t\underline{/\delta_t}$ and the phase current is $\overline{I}_s = I_s\underline{/\phi_s}$. Remember that the per phase equivalent circuit is based on a wye connected machine, and the voltage used in the equation is the line to neutral voltage. The angles are the voltage angle and current angle at the machine terminal.

The reactive power of the stator winding is

$$Q_{3\phi} = 3V_tI_S\sin(\delta_t - \phi_s) \tag{5.28}$$

The apparent power is

$$S_{3\phi} = 3V_tI_S \tag{5.29}$$

The direction of real and reactive power is the same as the direction chosen for the reference direction of the current. With the current direction shown in figure 5.6, positive power will flow out of the machine. This happens during generator operation, when mechanical power is supplied to the machine through the shaft. During motoring operation, the computed real power $P_{3\phi}$ would be negative when computing the power flow from figure 5.6. When studying motoring operation, it is common to use the direction of current flow shown in figure 5.8. In this case, equations (5.26) and (5.27) describe the real and reactive electrical power *entering* the machine.

Example 5.7. The per phase diagram of a three phase synchronous machine is shown below. Determine the phasor current flow in the machine, and the magnitude and direction of real and reactive power flow.

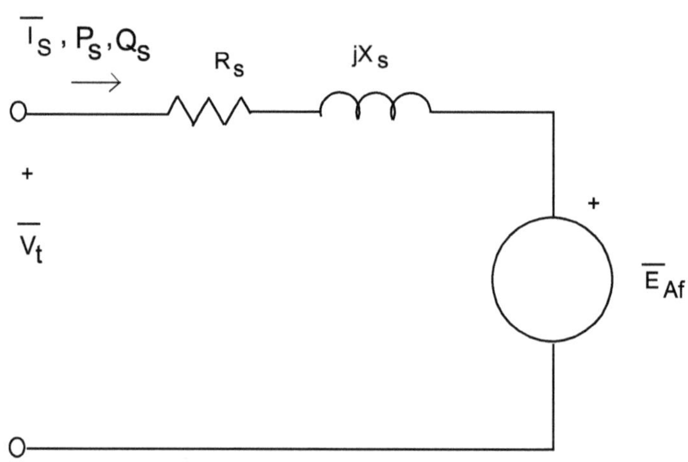

Figure 5.8. Synchronous machine with motor convention.

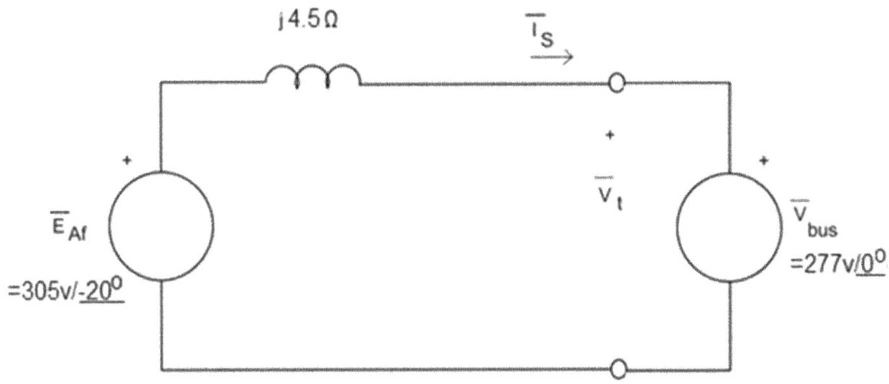

Solution. The current in the machine is

$$\bar{I}_S = \frac{\bar{E}_{Af} - \bar{V}_{bus}}{\bar{Z}_{synch}} = \frac{305\text{ V}\underline{/-20^\circ} - 277\text{ V}\underline{/0^\circ}}{j4.5\Omega} = 47.15\text{ A}\underline{/-177.4^\circ}$$

From equation (5.27), the power flow OUT of the machine terminals is

$$P_{3\phi} = 3V_tI_S\cos(\delta_t - \phi_t) = 3\cdot 277\text{ V}\cdot 47.15\text{ A}\cdot\cos(0^\circ - (-177.4^\circ)) = -39.14\text{ kW}$$

This is the same as +39.15 kW flowing into the machine, so this machine is motoring.

The reactive power of the machine is

$$Q_{3\phi} = 3V_tI_S\sin(\delta_t - \phi_t) = 3\cdot 277\text{ V}\cdot 47.15\text{ A}\cdot\sin(0^\circ - (-177.4^\circ)) = +1.78\text{ kVARs}$$

As these VARs are positive in this generator convention calculation, 1.78 kVAR are flowing out of the machine.

It is instructive to repeat this example with the motor convention. The result would be the same—39.14 kw flowing into the machine and 1.78 kVAR flowing out of the machine (−1.78 kVAR flowing into the machine), but the current angle and sign on the calculate P and Q values will be different.

5.6 Mechanical power and torque—generator convention

Equation (5.26) is the power delivered by the generator to an electrical load. The machine is acting as a generator in this case. In the generator, mechanical power is delivered to the machine, converted to electrical power, and delivered to the load. Machine losses reduce the power delivered.

Figure 5.9 shows a mechanical diagram of a turbine-generator rotor. In practice, the generator can be driven by a steam, gas, wind or hydraulic turbine, a piston engine, or other prime mover. The mechanical power delivered to the shaft by the

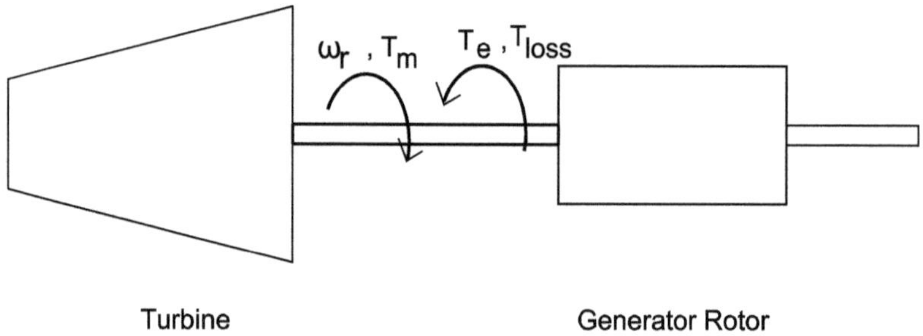

Figure 5.9. Turbine-generator rotor diagram showing speed and torque direction.

turbine is matched by the power taken from the shaft by the generator. In rotational systems,

$$P_{\text{shaft}} = T_{\text{m}}\omega_{\text{r}} \tag{5.30}$$

As shown in the figure, the torque of the generator opposes motion in generator mode. When the supplied torque matches the generator torque, the system is in balance and operating at a constant speed. When the torques are not in balance, the turbo-generator shaft will either accelerate or decelerate,

$$J\frac{d\omega_{\text{r}}}{dt} = T_{\text{m}} - T_{\text{e}} - T_{\text{loss}} \tag{5.31}$$

J is the moment of inertia of the rotor. T_{loss} represents the mechanical losses in the generator, and will oppose the shaft rotation.

Machine losses reduce the power delivered. The generator losses fall into five general categories.

(a) Mechanical loss—friction and windage.
(b) Core loss—eddy current and hysteresis.
(c) Stator copper loss and stray losses.
(d) Field copper loss.
(e) Any power required to cool the machine.

The mechanical loss is due to friction and windage. Friction losses are primarily due to bearing losses. The windage loss is aerodynamic loss as the rotor spins in air or other gas. In some designs, the rotor includes a set of fan blades that pull air through the machine to provide cooling. The magnetic core loss, due to hysteresis and eddy currents in the machine, are a function of flux density and frequency. As these losses are roughly linear over small ranges of shaft speed, they are often modeled as additional mechanical losses, and included in the term T_{loss}. This provides suitable accuracy for the model, and simplifies the electrical calculations.

Together, these losses reduce the torque supplied by the turbine. The remaining torque T_{e} opposes the shaft rotation in a generator, and supplies the power converted to electrical power by the generator.

$$P_{\text{conv}} = T_{\text{e}}\omega_{\text{r}} \tag{5.32}$$

The converted power appears electrically on the stator armature windings. This electrical power is reduced by the stator copper loss and stray load loss, which together are represented by the stator winding resistance R_s in the per phase equivalent circuit of the generator. The field power and any power required to cool the machine apart from shaft mounted fans generally come from sources other than the turbine generator shaft, and are included in efficiency formula as additional input powers. The field winding loss component is not present in permanent magnet synchronous machines. Wound field machines have the advantage of allowing the field level to be adjusted to accommodate various loading conditions. Permanent magnets provide a set field level, but have the advantage of simplicity as they don't require a field winding on the rotor. Both have practical application.

The electromagnetic field losses are an electrical loss, as shown in figure 5.10. However, it is often convenient to model them as a mechanical loss, for calculation purposes. This does have some effect on the value of the converted power and the developed torque T_e, but does not affect the electrical or mechanical terminal values.

Figure 5.10 shows the flow of power in a synchronous generator with wound field. The mechanical input power P_{shaft} is given by equation (5.30), and the electrical output power $P_{3\phi}$ is calculated by equation (5.27). The efficiency of the generator is

$$\eta = \frac{P_{out}}{P_{in}} = \frac{P_{3\phi}}{P_{shaft} + P_{aux}} \tag{5.33}$$

This value is often multiplied by 100 to give efficiency in per cent.

In a synchronous motor, the power flow is reversed. The electrical power is input to the stator of the motor, and the shaft power is the motor output. The power flow diagram for a synchronous motor is shown in figure 5.11.

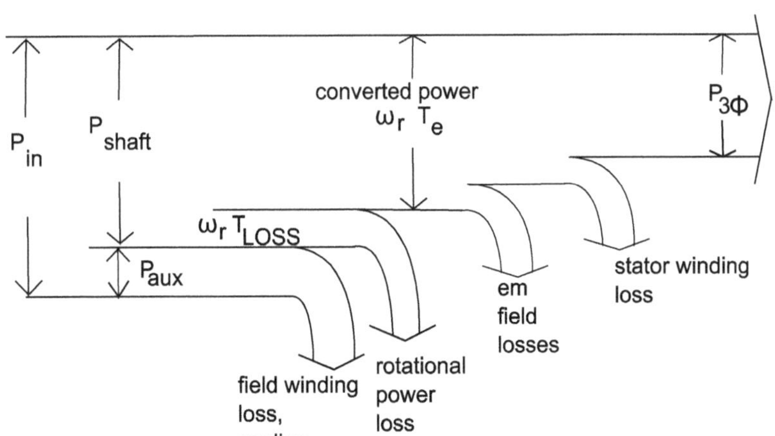

Figure 5.10. Power flow diagram for synchronous generator.

5.7 Distributed windings and salient pole designs

5.7.1 Distributed windings in round rotor machines

In previous sections, it was assumed that the flux density was distributed sinusoidally around the air gap. In fact, it is a challenge to design a machine with sinusoidal flux distribution.

Consider the two pole round rotor machine shown in figure 5.3. Assume that the reluctance of the iron path is negligible, and that the air gap length is constant. The single A phase winding creates a constant magnetomotive force (MMF) across the air gap to the right of the coil, and the same magnitude MMF in the opposite direction on the left half of the machine. As a result, the flux density around the air gap will be constant around each half of the air gap, as shown in figure 5.12. A 'straightened out' view of the machine is shown on the right in this figure, to allow a plot of the flux density to be shown below the physical location of the machine.

Round rotor machines use distributed windings to approach a sinusoidally distributed flux density. Figure 5.13 shows a round rotor machine with the A phase

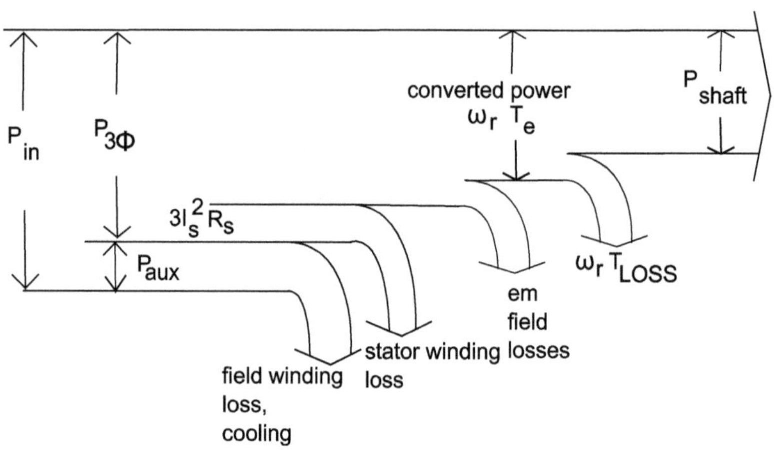

Figure 5.11. Power flow diagram for a synchronous motor.

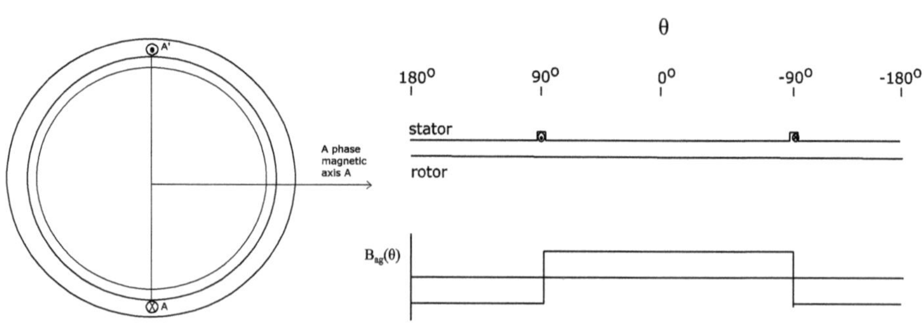

Figure 5.12. Flux density waveform for a cylindrical machine with a concentrated winding.

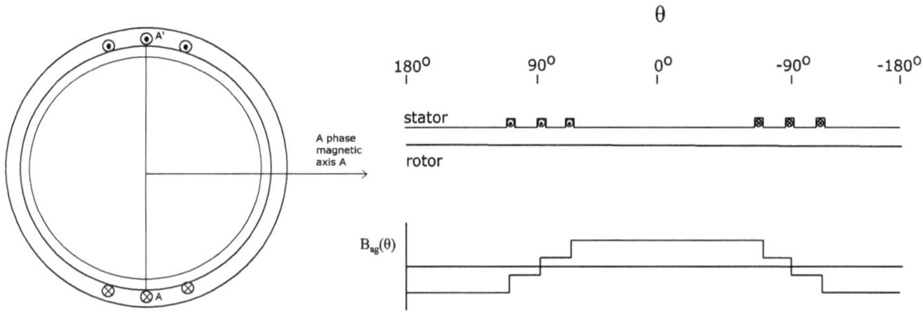

Figure 5.13. Distributed winding with three coils separated by 15°.

winding distributed in three coils in three slots. While each coil creates a square wave flux density, the sum effect of the three coils improves the flux density distribution.

Machine designers balance the slot number, number of coils, coil distribution, pitch and other factors to develop round rotor synchronous and induction machines with suitably sinusoidal flux density distributions.

5.7.2 Salient pole machines

Salient pole machines have a variable air gap length. The windings are generally concentrated. A conceptual diagram of a four pole salient pole wound field machine is shown in figure 5.5. In this machine, the stator and rotor pole faces are shaded in order to approach a sinusoidally distributed flux density wave form. In this case, the air gap along the field axis (direct or *d* axis) is smaller than the air gap between the field poles (90° away electrically, and often referred to as the quadrature or *q* axis).

In variable air gap machines, the stator current is resolved along the *d* and the *q* axes, and the *d* and *q* currents see different synchronous inductances due to the different air gap lengths. The per equivalent circuit shown in figure 5.6 has a single value for synchronous inductance L_S, so is not particularly accurate for salient pole machines.

It is interesting to note that a salient rotor itself will interact with armature windings on the stator to produce torque. In most cases, salient pole machines combine field flux and saliency to produce torque. However, a class of machines called synchronous reluctance machines rely solely on this saliency torque to operate.

5.7.3 Permanent magnet machines

There are two basic designs for permanent magnet synchronous machines—rotors with surface magnets and rotors with embedded magnets.

Figure 5.14(a) shows a diagram for a rotor with surface mounted permanent magnets. As the relative permeability of the permanent magnets approaches $\mu_r = 1$, the combined permanent magnet and air gap length is essentially constant around the motor.

Figure 5.14(b) shows the basic layout for an embedded magnet rotor. In the case shown, the rotor is round, with buried magnets producing flux on the *d* axis, and high permeability magnet steel on the *q* axis. In this case, the apparent air gap length on the *d* axis is large while the gap length on the *q* axis equals the actual air gap. As a

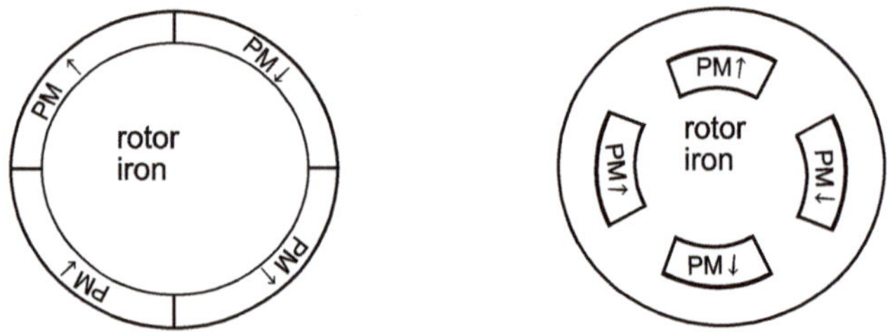

(a) Surface Mount Design (b) Interior Magnet Design

Figure 5.14. Cross sectional views of surface mount and interior permanent magnet machine rotors.

result, the d axis reluctance is large relative to the q axis reluctance. This causes the d axis inductance to be smaller than the q axis inductance.

5.8 Salient pole machines

The previous sections considered the modeling and performance of round rotor synchronous machines.

Salient pole machines are a second type of synchronous machines. In salient pole machines, the effective air gap at the field poles is different than the air gap on the pole face. In these machines, the field magnetic axis is often referred to as the direct or D axis. The quadrature or Q axis is 90 electrical degrees away from the direct axis. Figure 5.15 shows these two axes in a three phase two pole machine.

Equation (5.5) shows that the peak value of the field flux will lie on the field, or D axis. The field flux induces the internal voltage E_{Af}. Equation (5.7) shows that this voltage leads the D axis by 90°—while the field flux peaks at angle θ, the induced voltage peaks at an angle of $\theta + 90°$.

It is useful to define space vectors that have the magnitude of the given value, and the direction pointing to the peak value of the wave. Figure 5.16 shows this same two pole machine with field flux Φ_{f} lying on the D axis (at angle θ), and the internal voltage E_{Af} lying on the Q axis.

As is the case with round rotor generators, current flows when the stator windings are connected to a load or source. A current space vector is also shown in figure 5.16, with the case of the current lagging the internal voltage E_{Af} by angle ϕ. This current in turn will create the rotating stator flux wave, which interacts with the field flux to create torque. In the round rotor case, this flux was proportional to current, and at the same angle as the stator current.

In salient pole machines, the stator current needs to be separated into two components, one along the D axis, and the other on the Q axis. These two current components will create different levels of flux per amp, due to the different air gap lengths. The current component on the axis with the smaller air gap will create more flux per amp than the current on the axis with the larger air gap. The result is that the stator flux wave is at a different angle to the space vector stator current. This difference in flux created per amp corresponds to different inductances on the two axes.

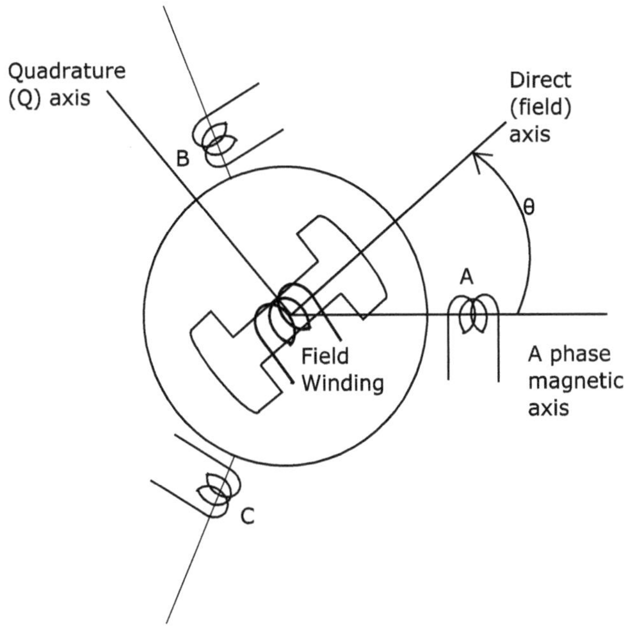

Figure 5.15. Three phase two pole salient pole machine showing the direct and quadrature axes.

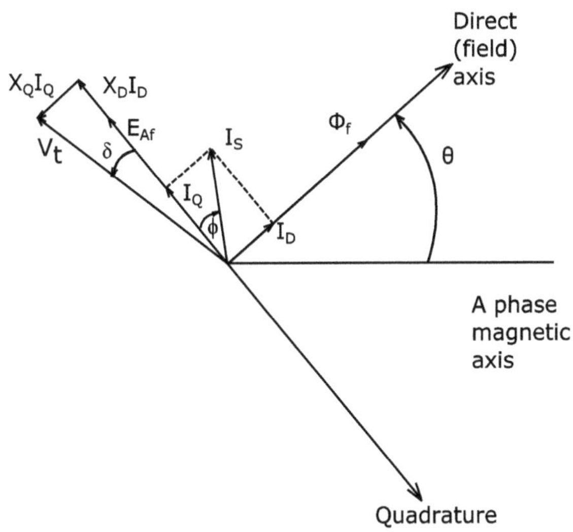

Figure 5.16. Two pole machine showing the stator current space vector and its D and Q components.

In wound field machines, the D axis air gap is always smaller than the Q axis air gap. We will define the D axis inductance L_D as the inductance that the D axis current sees. Similarly, the Q axis inductance L_Q is the inductance that the Q axis current component will see. Both of these include the stator leakage inductance, as is also the case with the synchronous inductance in a round rotor machine. In a wound field salient pole machine, L_D will be larger than L_Q due to the smaller air gap on the D axis. The current I_D will create a voltage on the Q axis, equal to $X_D I_D$. Similarly, the current I_Q will create a voltage $X_Q I_Q$ in the negative direction on the D axis. Both induced voltages lead the respective current by 90°, the same as in the case of a round rotor machine. The difference is that the synchronous reactance is different for the D and the Q axes in the salient pole case.

With the stator winding resistance neglected, the resulting space vector relationships are:

$$\begin{aligned} V_Q &= E_{\mathrm{Af}} + X_D I_D \\ V_D &= -X_Q I_Q \end{aligned} \tag{5.34}$$

From these equations, the power entering the motor terminals can be found to be

$$P_{\mathrm{M}} = \frac{3V_{\mathrm{t}}E_{\mathrm{Af}}}{\omega_{\mathrm{e}}L_{\mathrm{D}}}\sin\delta + \frac{3V_{\mathrm{t}}^2}{2\omega_{\mathrm{e}}}\left(\frac{L_D - L_Q}{L_D L_Q}\right)\sin 2\delta \tag{5.35}$$

In this equation, V_{t} is the RMS line to neutral generator terminal voltage, ω_{e} is the generator electrical frequency in radians/second, δ is the angle by which the terminal voltage leads the Q axis, and L_D and L_Q are the D and Q axis synchronous inductances, respectively. The corresponding developed torque can be found by dividing equation (5.35) by the shaft speed.

In equation (5.35), the first term is identical to the power-angle term in a round rotor machine. The second term depends on the machine saliency, as exhibited by the difference values of L_D and L_Q. This torque component is often referred to as the reluctance torque, or the variable reluctance torque. Salient pole synchronous machines develop both torque components. Also, there is a class of machine with salient pole rotors and no field winding or permanent magnets. These machines develop only the second torque component, and are referred to as reluctance machines or synchronous reluctance machines.

5.9 Summary

This chapter covers the basic characteristics of AC machines, particularly the synchronous machine. AC machines are widely used for both generation and motoring, and the next chapters cover AC machine operation.

Questions

1. Professor Riaz at the University of Minnesota developed several animations showing machine flux density distribution. These can be found at: http://people.ece.umn.edu/users/riaz/animations/listanimations.html
   a. Study animations 9–12 and 15, and comment on the nature of the stationary and rotating flux waves.
   b. Animations 13, 14 and 16 show space vectors. These vectors show the magnitude and direction of the flux density wave. These can also represent the currents that generate the flux wave.

2. a. Do an internet search for synchronous machine rotors. Salient pole rotors will have a number of prominent pole faces with windings around the pole. Describe a salient pole rotor that you find an image for.
   b. Do a search for a synchronous machine stator. Select an image with the rotor removed, and describe the coils and slots of the stator.
   c. Search for an induction motor exploded view. Select one the clearly shows the squirrel cage rotor. Name the various components of the rotor assembly.
3. In practice, machine efficiency is often calculated as $\frac{P_{out}}{P_{out}+P_{Loss}}$. Write this equation for the generator operation, showing each loss component individually.
4. Repeat question 2, but for motor operation.

Problems

1. A three phase two pole 60 Hz wye connected synchronous generator is rated at 2.50 MVA, 4.16/2.4 kV. The generator has a stator resistance of $R_s = 0.15\ \Omega$ and a synchronous reactance of $X_s = 4.0\ \Omega$.
   a. Find the rated current of the generator
   b. The generator is operating at rated terminal voltage with rated current. The current is lagging the line to neutral voltage by 25°. Find the internal voltage $\overline{E}_{Af}$ of the generator.
   c. Find the converted power and the developed torque T_e.
2. A three phase 60 Hz synchronous motor has four poles. The motor rated voltage is 460 V/266 V wye. The motor rated stator current is 175 A. The per phase diagram of the motor is shown below. The motor is operating with a terminal voltage magnitude of $V_t = 270$ V with a frequency of 60 Hz. The current magnitude $I_s = 175$ A. I_s is in phase with V_t.
   a. Calculate the magnitude and angle of the internal voltage $\overline{E}_{Af}$.
   b. The internal voltage magnitude $E_{Af} = 0.075\omega_e I_f$, where ω_e is the electrical frequency in radians per second, and I_f is the field current. Find the field current for this operating point.
   c. The converted power for the motor is the real power entering the internal voltage source E_{af}. Find the converted power for this motor.
   d. From the converted power and the rotor speed ω_r, find the developed torque of the motor.

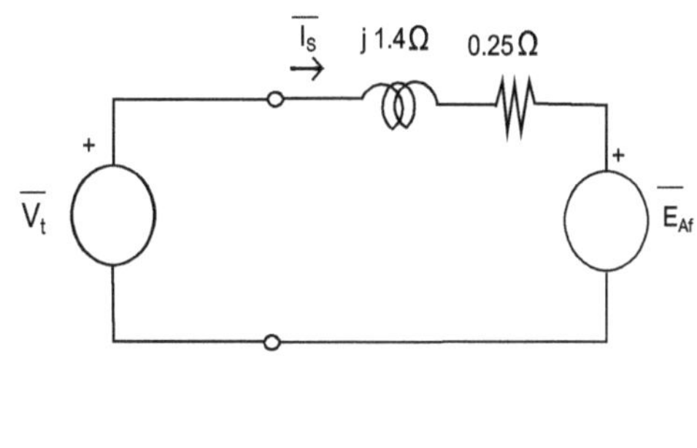

3. The motor of problem 2 is operating with a developed torque of $T_e = 500$ N m and a field current of $I_f = 20$ A. The electrical frequency is 60 Hz.
   a. Determine the internal voltage E_{Af}.
   b. Calculate the converted power $P_{conv.}$
   c. Assume that the stator current is in phase with the internal voltage. Determine the stator current magnitude $I_{s.}$
   d. From this, calculate the motor terminal phasor voltage $\overline{V}_t$.
   e. Calculate the terminal real power P_{in} and reactive power $Q_{in.}$
4. Repeat example 5.7, using the motor convention instead of the generator convention.
5. A four pole 60 Hz synchronous generator has 145 MW flowing from its terminals. It has a stator copper loss of 1.3 MW, friction and windage loss of 0.8 MW, electromagnetic losses of 1.2 MW and field winding and cooling losses of 2.1 MW. The generator is operating with electrical frequency of 60 Hz.
   a. Determine the rotational speed of the generator in RPM and in radians per second
   b. Using the power flow diagram of figure 5.10, determine the converted power of the machine, and the developed torque T_e
   c. Determine the shaft power P_{shaft} delivered by the turbine, and the corresponding torque T_m. Compare the difference between T_e and T_m.
   d. Determine the overall efficiency of the generator at this operating point.

IOP Publishing

Electromechanical Machinery Theory and Performance

Thomas Ortmeyer

Chapter 6

Synchronous machine performance

The theory and basic characteristics of synchronous machines was presented in chapter 5. In this chapter, these machine characteristics will be used to study the performance and application of these machines.

6.1 Synchronous generators

Three phase synchronous generators are the primary means of generating electric power today. They range in size from a few kVA to over 1000 MVA. These machines carry a number of ratings that describe their capability and limits of usage.

Synchronous generators can be expected to operate successfully for long periods of time—25 years of successful operation is common for these machines, and lifetimes regularly exceed this age. In order to achieve these lifetimes successfully, they must be installed properly, be operated within their capabilities, and be properly maintained.

A primary set of ratings for a synchronous generator describe its steady state electrical characteristics. A partial set of typical ratings is given in table 6.1. The generator must be operated according to these ratings in order for the machine to perform as expected and to avoid shortening the lifetime of the generator. The governing standard for the generator (typically an IEEE/ANSI/NEMA or IEC document) describes the meanings of the ratings. For example, the continuous volt–amp (VA) rating should not be exceeded for any significant period of time, and the machine can be operated with VA outputs significantly lower than rated. On the other hand, the machine should be operated within a range of the rated voltage and frequency—perhaps $\pm 5\%$ for voltage and $\pm 2\%$ for frequency. Also shown in the table are machine characteristics that allow the owner/operator to properly operate the machine. The ambient temperature and altitude above sea level, on the other hand, define the range of conditions for which other machine ratings apply—the generator, for example, could be operated at ambient temperatures above 40 °C, but at a reduced volt–amp limit.

doi:10.1088/978-0-7503-1662-0ch6

Table 6.1. A selected portion of the ratings and characteristics of a typical synchronous generator.

Ratings		Characteristics	
Governing standard	Maximum overspeed	Inertia constant	Rated efficiency
Speed	Rated power factor	Direct axis inductance (unsaturated)	Stator winding resistance
Frequency	Short term overload rating	Direct axis inductance (saturated)	Field winding resistance
Voltage	Temperature rise	Quadrature axis inductance (unsaturated)	Maximum short circuit current
Continuous VA output	Ambient temperature	Quadrature axis inductance (saturated)	
Direction of rotation	Maximum altitude above sea level	Voltage and current unbalance limitations	
Insulation class	Field voltage		

The combination of frequency and shaft speed give the number of poles of the generator

$$\text{poles} = \frac{120 f_{\text{rated}}}{\text{rpm}_{\text{rated}}} \tag{6.1}$$

Similarly, the current ratings of the stator armature can be determined from the other ratings of the machine.

The voltage ratings for three phase machines are stated differently for delta connected and wye connected generators. In a delta connected generator, the phase windings are connected from line to line, so the winding voltage and line to line voltages are the same. The voltage rating is stated as a single number—e.g. 4160 V. This voltage is the line to line voltage of the supplying system and also the voltage across each winding of the delta. With wye connected windings, the phases are connected from line to neutral. The line to line voltages are larger by a factor of $\sqrt{3}$. The voltages are stated as, for example, 4160 V/2400 V. In this case, 4160 V is the rated line–line voltage, and 2400 V is the rated voltage of each phase winding, which is also the line to neutral voltage. This method of stating voltage follows ANSI Standard C84.1-1989. (ANSI is the American National Standards Institute.) This standard also states that the system voltage is defined as the RMS line to line voltage.

The volt amp rating is the rated value of three phase volt–amps at rated voltage. The maximum stator winding current capability can be found from this volt–amp rating, with certain exceptions defined by the applicable standard. The generator itself can supply this level of volt–amps at power factors above the rated power factor. The actual real power limitation of a turbine generator system will be set by the turbine and prime mover driving that turbine. The generator cannot be relied on to supply the rated volt–amps at power factors less than rated power factor.

In particular, field winding current limits can become the limiting factor at low lagging power factors. The generator may be capable of operating in the steady state at power factors below rated, but not at rated volt–amps in this case. While operation of the generator may be permitted above the stated altitude, the generator may need to be derated or otherwise limited at those altitudes.

6.2 Determining synchronous machine parameters by test

The synchronous machine parameters are generally determined by a series of standard tests. The steady state parameters are determined through two tests, the open circuit test and the short circuit test.

Open circuit test: In the open circuit test, the generator is driven at rated speed by a mechanical source, and with the stator winding open circuited. Field current is increased, and the stator terminal voltage is measured. The resulting no load excitation curve was shown in figure 5.7.

In this figure, the point where the stator voltage equals rated voltage is marked. As you can see, there is some level of saturation at this point. We define the generator field constant K_e as

$$\omega_e K_e = \frac{V_{s(R)}}{I_F} = \frac{V_{ag(R)}}{I_F} = \frac{E_{Af(R)}}{I_F} \tag{6.2}$$

As there is no current, the internal voltage $\overline{E}_{Af}$, air gap voltage $\overline{V}_{ag}$ and terminal voltage $\overline{V}_t$ are all equal in this test. As can be seen from figure 5.7, the excitation curve is a non-linear function of field current. It is therefore necessary to take this measurement at rated voltage, which is at the nominal operating point for the machine.

Note that, under load, the generator air gap flux is a function of both the field current and the stator current. In a round rotor machine, the level of saturation is a function of this resultant air gap flux. The voltage induced on the stator winding by this resultant flux is called the air gap voltage. A per phase equivalent circuit of the generator is shown in figure 6.1. The internal voltage $\overline{E}_{Af}$ is the induced by the field flux, and the air gap voltage $\overline{V}_{ag}$ is the voltage induced by the resultant air gap flux. These two voltages are separated by the stator winding internal reactance X_{SW} in the

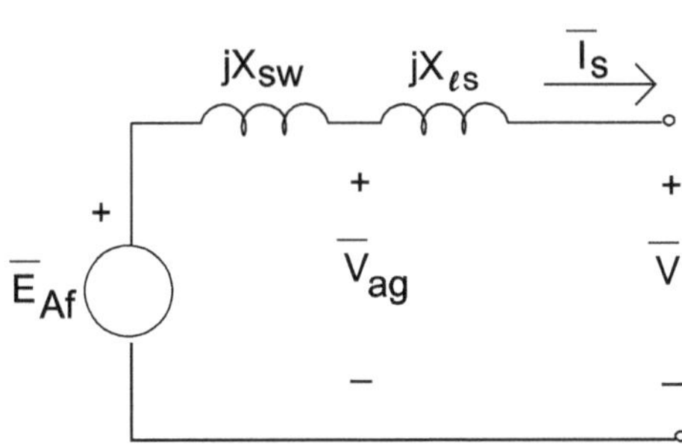

Figure 6.1. Synchronous generator per phase equivalent circuit showing induced voltage, air gap voltage, and terminal voltage.

per phase equivalent circuit. The voltage drop across X_{SW} corresponds to the flux linkage created by the stator current $L_{SW}i_S$, when balanced three phase currents are flowing. The stator winding also has leakage inductance $L_{\ell s}$, and the difference between the air gap voltage and the terminal voltage is the voltage drop across this inductance. In steady state sinusoidal conditions, the voltage across each inductor is proportional to the time derivative of the current—this results in an inductive impedance of the frequency in radians per second times the inductance. The phase shift caused by the derivative is reflected by the term j, which represents a shift of 90°.

In practice, the leakage inductance is significantly smaller than the stator inductance L_{SW}, and the air gap voltage is not much different than the terminal voltage. When the terminal voltage varies over a small range, the air gap voltage variation will also be small. It can then be assumed that the field constant K_e is indeed constant, as the level of saturation changes only slightly.

In figure 5.7, the voltage-field current curve is a straight line at the lower excitation levels. This is because iron core saturation has not begun at this excitation level. This straight line portion of the curve is called the air gap line. The slope of this curve is the unsaturated field constant,

$$\omega_e K_{e(unsat)} = \text{slope of air gap line} \tag{6.3}$$

This value of the constant would be used when the machine is operated at low excitation levels.

The second set of measurements in the no load test is the shaft power that it takes to drive the machine at no load. This should first be measured with the generator unexcited, $I_F = 0$. The resulting power will be the rotational loss. The loss should next be measured at rated terminal voltage. This loss now will include both the rotational loss and the no load iron loss.

Both these losses can be considered to be constant in synchronous generators that operate over a small range of frequency and terminal voltage.

Short circuit test: The total generator stator phase inductance is determined by the short circuit test. In this test, the generator is rotated at rated speed. The generator stator windings are short circuited, and the field current is raised slowly until rated stator current is achieved. Typically, multiple points are taken, and the short circuit current as a function of field current is plotted. This will, approximately, be a straight line, as the magnetic core is not saturated under this condition.

As the stator air gap voltage will be low in this test, the per phase internal voltage can be determined as

$$E_{Af(sc)} = \omega_e K_{e(unsat)} I_f \tag{6.4}$$

Here, I_f is the field current of the machine. The stator resistance can often be neglected. If this is the case, the total per phase reactance of the generator is

$$X_S = \omega_e(L_{SW} + L_{\ell s}) = \frac{E_{Af(sc)}}{I_{rated}} \tag{6.5}$$

The DC resistance of the stator windings can be determined by applying a DC voltage across each of the phases, and measuring the current flow. This resistance should be adjusted to reflect the operating temperature of the winding, if the test is taken at some other temperature. It should also be adjusted slightly upward for 60 Hz currents due to the skin effect, winding temperature and stray load loss.

Example 6.1. Synchronous generator open circuit and short circuit tests.

A three phase, four pole 60 Hz round rotor synchronous generator is tested to determine the machine constants. The generator windings are wye connected, and the rated voltage of the winding is 4160/2400 wye volts. The preliminary data for the test is shown in the table below. Resistance data is adjusted for temperature and skin effect. The generator rating is 2.5 MVA.

Field resistance	Armature resistance	Rotational loss, no excitation	Rotational loss, rated open circuit voltage	Field current at rated open circuit voltage	Field current at rated short circuit voltage
1.05 ohms	0.052 ohms	31.4 kW	54.2 kW	26.3 A	34.5 A

The rated speed of the generator is

$$n_{\text{rated}} = \frac{120}{\text{poles}} f_{\text{e(rated)}} = 1800\,\text{rpm}$$

The rated stator current of the generator is

$$I_{\text{srated}} = \frac{S_{\text{rated}}}{3V_{\text{LNrated}}} = \frac{2.5 \cdot 10^6\,\text{VA}}{3 \cdot 2400\,\text{V}} = 347\,\text{A}$$

Open circuit test: The open circuit test is performed with the generator spinning at 1800 rpm, and the field current increased in steps until the terminal voltage reaches rated voltage. Figure 6.2 shows the generator open circuit curve taken at a rotor speed of 1800 rpm.

The open circuit curve shows the saturation curve and the air gap line. From this curve, at rated voltage of 2400 V line-neutral, the field current is $I_f = 26.3$ A. The saturated generator constant is then

$$\omega_e K_e = \frac{2400\,\text{V}}{26.3\,\text{A}} = 91.25.$$

The internal generator voltage is then

$$E_{\text{Af}} = 91.25\, I_f$$

This equation is valid for operating conditions where the generator air gap voltage is approximately equal to the rated voltage.

The air gap line is also shown in this figure. From this line, the relationship between E_{AF} and field current when the machine is unsaturated is

$$\omega_e K_{\text{e(unsat)}} = \frac{2400\,\text{V}}{25.4\,\text{A}} = 94.5$$

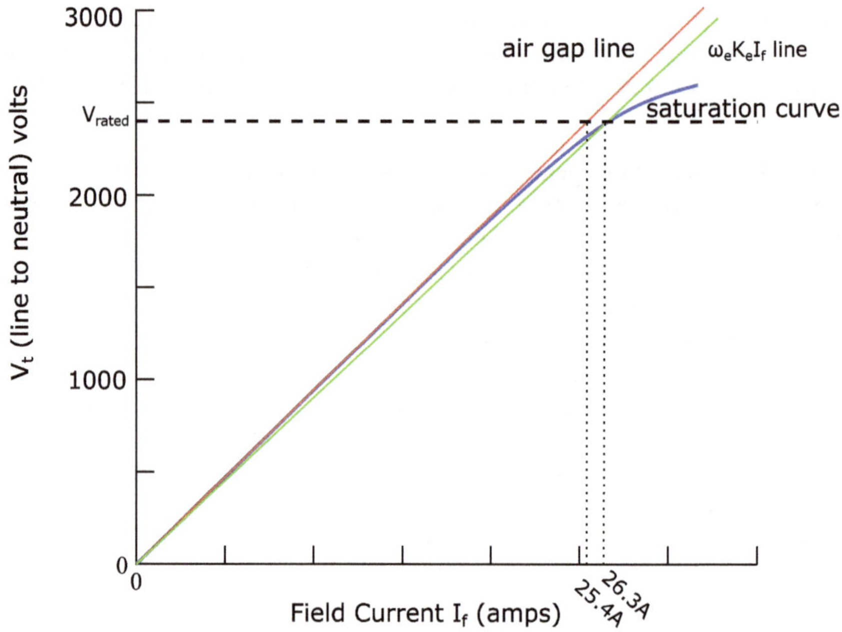

Figure 6.2. Generator open circuit curve, including air gap line.

Short circuit test: The short circuit test is conducted with the generator spinning at 1800 rpm, and the three phase stator winding short circuited. The field current is increased in steps until rated current is reached.

The short circuit test result is shown in figure 6.3. This data is used to find the synchronous reactance of the machine.

The synchronous reactance X_s is the inductive reactance of the armature (phase) windings of the generator, as shown in figure 6.4. When the stator phases are short circuited, the synchronous reactance is

$$X_s = \frac{E_{AF}}{I_s}$$

In the short circuit test, the generator is operating with a low value of air gap voltage. The machine is therefore not saturated. The internal voltage E_{AF} is then determined from the air gap line, using the constant $\omega_e K_{e(unsat)}$. The short circuit current I_s is also a function of the field current in the short circuit test. The measured short circuit data are given in the table above. For the field current, the synchronous reactance is found to be

$$X_s = \frac{E_{AF}}{I_s} = \frac{\omega_e K_{e(unsat)} I_F}{I_s} = \frac{(94.5 \cdot 34.5)\text{V}}{347\text{ A}} = 9.39\ \Omega$$

Losses: The final step of identifying the steady state machine parameters is taken from the loss data from the open circuit test. At rated speed, it takes 31.4 kW to drive the machine when both the stator and field circuits are open. This loss is due to the mechanical friction and windage losses of the rotor. When this same test is run with the field current present to create rated open circuit voltage, these losses increase to 54.2 kW. This increase is due to the no load iron losses in the machine core. In this case, the iron loss is 22.8 kW. In most cases when analysing the generator performance, the friction and windage loss and the core loss are treated as a single mechanical loss. This assumption is valid when the machine is operating near rated speed with near rated air gap voltage.

6.3 Synchronous generator operation

Synchronous generators are operated in two basic modes—standalone and grid connected. In standalone operation, a single generator directly feeds a load or set of loads. In the simplest case, the loads can be represented as a balanced, three phase impedance connected directly at the generator terminals.

The full three phase diagram for a wye connected generator directly feeding a wye connected load is shown in figure 6.4, along with the per phase equivalent circuit. Note the terminology in the per phase equivalent circuit that the stator current is $\overline{I}_S$. In the corresponding three phase diagram, $\overline{I}_A = \overline{I}_S$, and the B and C phase currents have the same magnitude but lag by 120° and 240° respectively.

Terminal conditions known: In some cases, the generator terminal voltage and current are known—both magnitude and phase angle. The generator internal voltage can then be found from the per phase equivalent circuit,

$$\overline{E}_{Af} = \overline{V}_t + (R_s + jX_s)\overline{I}_S \tag{6.6}$$

Figure 6.5 shows the phasor diagram of this equation for three cases—lagging current, unity power factor, and leading current. The generator synchronous

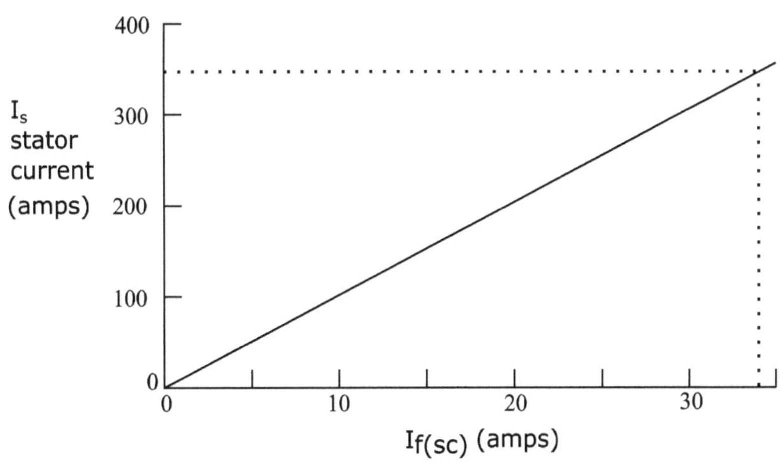

Figure 6.3. Generator short circuit curve.

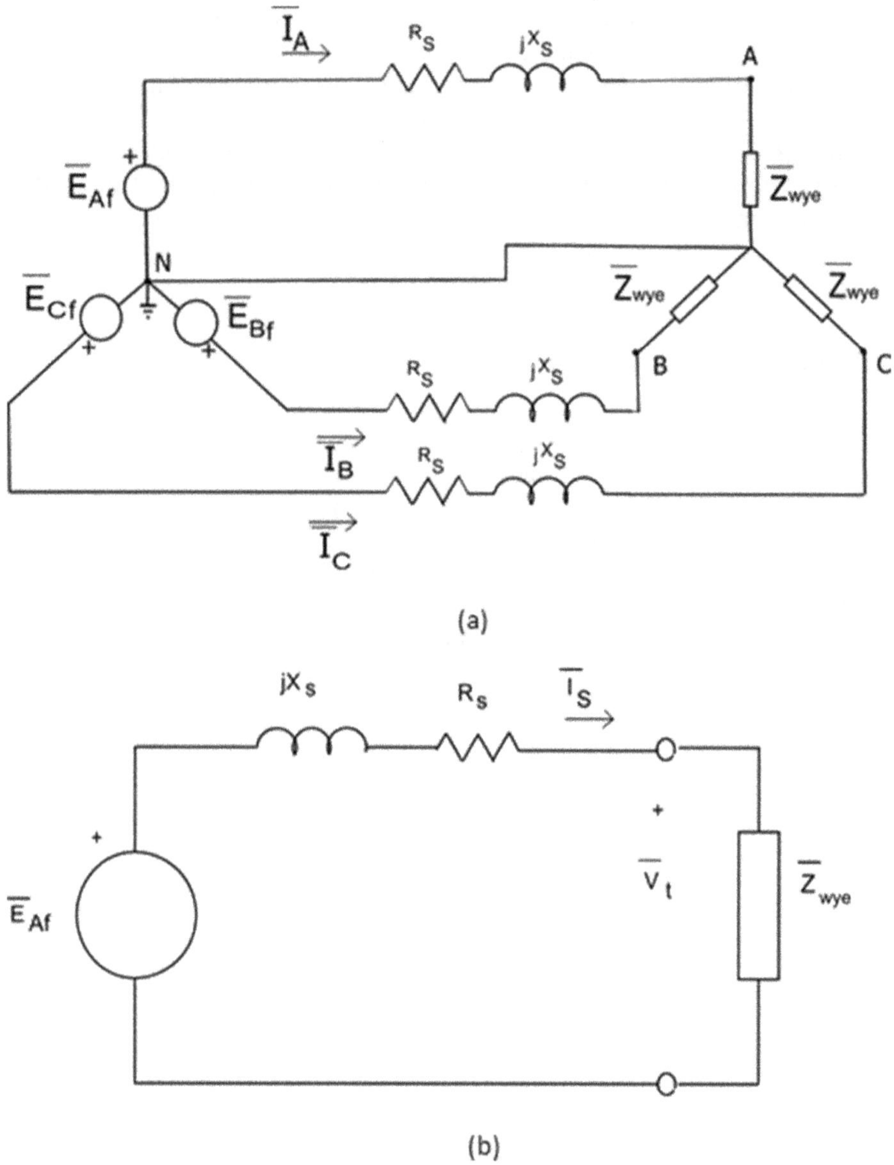

Figure 6.4. Equivalent circuits, synchronous generator feeding impedance load. (a) Three phase diagram. (b) Per phase equivalent circuit. Note the different terminology used in the per phase equivalent circuit.

reactance X_s is significantly larger than the phase resistance R_s, and this resistance is sometimes neglected, particularly for large generators. In the unity power factor case, the voltage drop across the synchronous reactance is perpendicular to the stator voltage. With lagging current, the voltage drop across the synchronous reactance is rotated to the right, so that the internal voltage E_{Af} is larger than in the unity power factor case. With leading current, the voltage drop across the

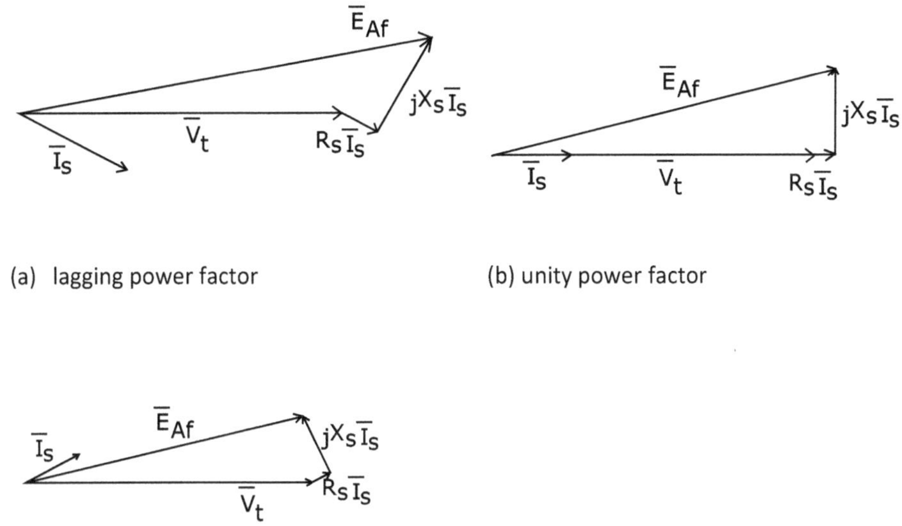

Figure 6.5. Phasor diagrams for synchronous generator.

synchronous reactance is rotated to the left. The result is that this tends to reduce the internal voltage E_{Af} from the unity power factor case. In fact, at some point, the magnitude of E_{Af} will become smaller than the terminal voltage—even though the real power flow is from the generator into the load.

When the terminal phasor voltage and current are known, the relationships between voltage, current, real power, and reactive power can be determined. It is important to remember that the voltages in the per phase equivalent circuit are line to neutral voltages.

$$\begin{aligned} S_{3\phi} &= 3\,V_t I_S \\ P_{3\phi} &= S_{3\phi}\cos\theta_t \\ Q_{3\phi} &= S_{3\phi}\sin\theta_t \end{aligned} \tag{6.7}$$

With power factor $= \frac{P_{3\phi}}{S_{3\phi}}$ where

$$\begin{aligned} \overline{V}_t &= V_t \underline{/\delta_t} \\ \overline{I}_S &= I_S \underline{/\phi_s} \end{aligned}$$

and

$$\theta_t = \delta_t - \phi_S$$

It is also necessary to know if the power factor is leading or lagging. The power factor is lagging when current lags voltage, which is equivalent to θ_t being positive. In this case, the VARs are positive, indicating the generator supplying VARs to the system.

When terminal conditions are known, the internal voltage $\overline{E}_{Af}$ can be determined. From the magnitude E_{Af}, the field current can be determined.

It is interesting to note how the magnitude of the internal voltage E_{Af} is influenced by the phase angle of the load current as well as the magnitude of the phase current. The voltage angle δ_t also tends to increase as the power factor goes from lagging to unity to leading.

Excitation control: As the synchronous reactance is relatively large, the terminal voltage would drop significantly if the internal voltage E_{Af} was fixed. It is important to keep the terminal voltage relatively constant as the load changes. To do this, the field current must be increased as the load increases, in order to keep the terminal voltage within its required bounds. The controller that does this is referred to as the excitation controller or the voltage regulator. A block diagram for a generator terminal voltage regulator is shown in figure 6.6. The voltage regulator compares the terminal voltage V_t with the desired terminal voltage V_t^*. The controller responds to the error signal, and adjusts the field voltage to bring the terminal voltage back toward its desired value. The field circuit of synchronous generators requires significant power, so that controller output signal must be amplified up to the level required by the generator field. The power for the field circuit can come from a separate small generator located on the turbine generator shaft, or from power taken from the power grid or generator terminals.

In most cases, a proportional controller is used in the excitation circuit. This controller provides a controllable level of droop in the terminal voltage. The term 'droop' is defined as the change in a regulated quantity of a device (such as generator voltage or speed) as the device loading is increased. This voltage regulator is designed so that the voltage terminal voltages drops in a controlled manner as the generator is loaded. This controlled voltage droop is generally substantially less than the terminal voltage drop that would occur with no control.

In the steady state, with a proportional controller, the controller/amplifier block will have a constant value K_c. The field voltage V_{f0} is used establish to establish a base level of field voltage for the generator. The field voltage V_f will then be

$$V_f = K_c(V_t^* - V_t) + V_{f0} \tag{6.8}$$

In practice, the voltage regulator will have additional components to improve the dynamic performance, and to limit the field voltage from going too high or too low.

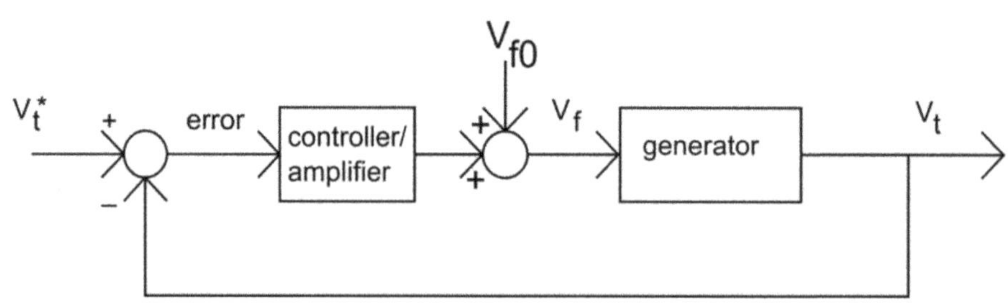

Figure 6.6. Conceptual block diagram of synchronous generator voltage regulator.

In the generator block, the steady state field current will be

$$I_f = \frac{V_f}{R_f} \tag{6.9}$$

The magnitude of the internal generator voltage is then, from equation (5.25),

$$E_{Af} = \omega_e K_e I_f \tag{6.10}$$

The generator terminal voltage can be determined from the per phase equivalent circuit of figure 6.4. With non-linear loads, such as electronic devices or induction motors, the relationship between the magnitude of the internal voltage E_{Af} and the terminal voltage V_t is complex and cannot be represented as a simple gain constant.

The voltage droop of a synchronous generator is defined as the drop in voltage from no load or light loading to full loading of the generator,

$$\text{voltage droop} = \frac{V_{t(\text{no load})} - V_{t(\text{full load})}}{V_{t(\text{no load})}} * 100\% \tag{6.11}$$

A voltage droop in the range of several percent as the generator goes from no load to full load is desirable for synchronous generator operation. This can only be achieved with the voltage regulation control that adjusts the generator field voltage to maintain the armature terminal voltage. A typical voltage regulation characteristic is shown in figure 6.7.

Voltage droop allows paralleled generators to cooperate effectively to regulate voltage and share VARs. The voltage droop provides some level of protection for the generator in response to short term overloads such as occur during induction motor starting.

Example 6.2. The synchronous machine of example 6.2 is operating with the 14 Ω wye connected load of that example. A second 14 Ω wye connected load is added in parallel with the first load. If the field current is unchanged, find the generator terminal voltage and the output power flow following the addition of this second load.

With the field current unchanged, the internal voltage of the generator will remain at 3168 V. Assume that the angle of the internal voltage is also unchanged, so $\overline{E}_{AF} = 3168\text{ V}\,\underline{/40.5^\circ}$. The per phase load impedance will be the parallel combination of two 14 Ω resistances, or 7 Ω.

The stator current will then be

$$\overline{I}_s = \frac{3168\text{ V}\,\underline{/40.5^\circ}}{0.052 + j12 + 7.0\,\Omega} = 227.6\text{ A}\,\underline{/-19.1^\circ}$$

The line to neutral terminal voltage is then

$$\overline{V}_t = \overline{Z}_{wye}\overline{I}_s = 7\,\Omega \cdot 227.6\text{ A}\,\underline{/-19.1^\circ} = 1593\text{ V}\,\underline{/-19.1^\circ}$$

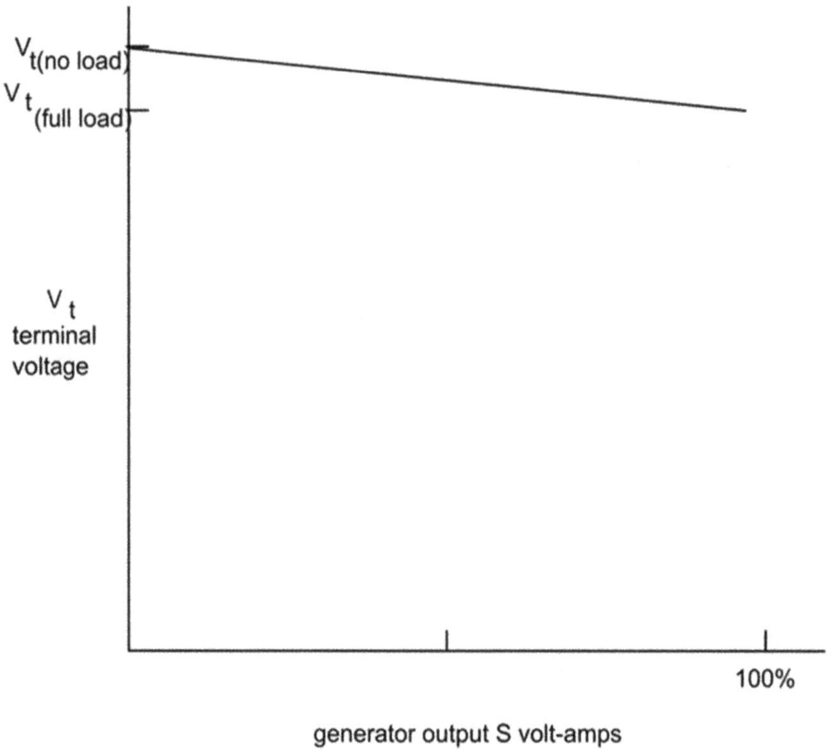

Figure 6.7. Generator voltage regulation characteristic.

The line–line terminal voltage magnitude is then $\sqrt{3} \cdot 1593\ \text{V} = 2759\ \text{V}$. The result of doubling the load on this generator is that the line to neutral voltage drops from 2400 V to 1593 V (the line to line voltage drops from 4160 V to 2759 V).

Clearly, this level of voltage drop is not acceptable, and the field voltage must be adjusted to bring the terminal voltage back into spec.

Example 6.3. A voltage regulator is added to the generator of example 6.2. This is a four pole machine operating at 60 Hz, so the generator shaft speed is 1800 rpm. The generator is initially loaded with a balanced wye connected load with a pure resistance of $\overline{Z}_{\text{wye}} = 14\ \Omega$. A second load of 14 Ω three phase load is then connected in parallel with the first load.

(a) Determine the generator phase current, internal voltage, field current and field voltage with the $\overline{Z}_{\text{wye}} = 14\ \Omega$ load.

Referring to figure 6.4(b), assume an angle of 0° for the terminal voltage, giving $\overline{V}_{\text{t}} = 2400\,\underline{/0^\circ}\text{V}$. The phase current is then

$$\overline{I}_{\text{s}} = \frac{2400\ \text{V}\,\underline{/0^\circ}}{14\ \Omega\,\underline{/0^\circ}} = 171.4\ \text{A}\,\underline{/0^\circ}$$

The internal voltage is then

$$\overline{E}_{AF} = \overline{V}_t + (R_s + jX_s)\overline{I}_s = 2400\ \text{V}\ \underline{/0^\circ} + (0.052 + j12\ \Omega)\cdot 171.4\ \text{A}$$
$$= 3168\ \text{V}\ \underline{/40.5^\circ}$$

The field current is found from the magnitude of the internal voltage,

$$I_f = \frac{E_{AF}}{\omega_e K_e} = \frac{3168\ \text{V}}{91.25} = 34.72\ \text{A}$$

Remember that the field circuit is DC. Since the field resistance is 1.05 Ω, the field voltage is

$$V_f = R_f I_f = 36.45\ \text{V}$$

(b) The second load is added, resulting in a $\overline{Z}_{wye} = 7\ \Omega$ load. If the voltage regulator brings the field voltage up to 50 V, find the voltage droop for this doubling of load.

With the new field voltage, the field current will become

$$I_f = \frac{V_f}{R_f} = 47.6\ \text{A}$$

Then the internal generator voltage will be

$$E_{Af} = \omega_e K_e I_f = 4345\ \text{V}$$

The stator current will be

$$\overline{I}_S = \frac{\overline{E}_{Af}}{R_s + jX_s + \overline{Z}_{wye}} = \frac{4345\text{V}\ \underline{/40.5^\circ}}{0.052\Omega + j12\ \Omega + 7\ \Omega} = 312.2\ \text{A}\ \underline{/-19.1^\circ}$$

The terminal voltage will then be

$$\overline{V}_t = \overline{I}_S \overline{Z}_{wye} = 2185\ \text{V}\ \underline{/-19.1^\circ}$$

The voltage droop will then be

$$\text{voltage droop} = \frac{V_{t(\text{light load})} - V_{t(\text{heavy load})}}{V_{t(\text{light load})}} * 100\%$$
$$= \frac{(2400 - 2185)}{2400}\cdot 100\% = 9.0\%$$

As we want the droop between two specific operating points, this equation was modified slightly from equation (6.11). As indicated by the lack of overbars on the voltages in this equation, droop is a real quantity calculated from the voltage magnitudes, rather than the complex phasor voltages. Finally, in this calculation it was assumed that $\overline{E}_{Af}$ maintained the same angle is in part (a). Note that this is an arbitrary decision, $\overline{E}_{Af}$ could have been taken to be 0° or any other value without affecting the voltage droop result

(c) Calculate the droop controller gain constant if $V_f^* = V_{f0} = 2400$ V.

With these given constants, the error voltage at light load is 0. At the heavy load, the error voltage is

$$\text{error} = 2400\text{ V} - 2185\text{ V} = 215\text{ V}$$

The output of the gain controller block is

$$\Delta V_f = 50\text{ V} - 36.45\text{ V} = 13.55\text{ V}$$

The controller/amplifier block steady state gain in then

$$G_{c/a} = \frac{13.55\text{ V}}{215\text{ V}} = 0.063$$

Example 6.4. For the generator and operating points of example 6.2, calculate the machine losses, output power, and efficiency.

For the first operating point, with $\overline{Z}_{wye} = 14\Omega$: from equation (6.7), the output power is

$$P_{out} = 3V_tI_s\cos(\delta_t - \phi_s) = 3 \cdot 2400 \cdot 171.4 \cdot \cos(0° - 0°)$$
$$= 1.234 \times 10^6\text{W} = 1.234\text{ MW}$$

The copper loss in the three stator phases is

$$P_{stator\ loss} = 3 \cdot 171.4\text{ A}^2 \cdot 0.052\ \Omega = 4584\text{ W}$$

The copper loss in the field winding is

$$P_{field\ loss} = 34.72\text{ A}^2 \cdot 1.05\ \Omega = 1265\text{ W}$$

The rotational loss and electromagnetic core loss from the open circuit test is

$$P_{rotational,\ core} = 54.2\text{ kW}$$

The total loss is then

$$P_{LOSS} = 4584\text{ W} + 1265\text{ W} + 54\,200\text{ W} = 60\,050\text{ W} = 60.05\text{ kW}$$

The generator efficiency at this operating point is then

$$\eta = \frac{P_{out}}{P_{out} + P_{loss}} = \frac{1.234\text{ MW}}{1.234\text{ MW} + 0.060\text{ MW}} = 0.954 \textit{ or } 95.4\%$$

For the second operating point ($\overline{Z}_{wye} = 7\Omega$), the

$$P_{out} = 3V_tI_s\cos(\delta_t - \phi_s) = 3 \cdot 2185 \cdot 312.2 \cdot \cos(-19.1° - (-19.1°)) = 2.046\text{ MW}$$

The generator losses are

$$P_{\text{loss}} = 47.6^2 \cdot 1.05 + 3 \cdot 312.2^2 \cdot 0.052 + 54\,200 = 71\,780 \text{ W}$$

The generator efficiency at this operating point is then

$$\eta = \frac{2.046 \text{ MW}}{2.046 \text{ MW} + 0.072 \text{ MW}} = 0.966 \text{ or } 96.6\%$$

It is also interesting to note that adding the second 14 Ω load resulted in a 66% increase in the output power. This is due to the voltage droop of the controller. If the voltage was held constant, a doubling of the output power could be expected for these impedance type loads.

Speed control: In a synchronous generator, the mechanical prime mover (energy source) must deliver sufficient power to supply the load plus the generator losses, as shown in figure 5.10. Generators are typically driven by steam, gas, hydraulic or wind turbines or piston engines. Power is delivered to the turbine by the energy source. The equation of motion for the turbine generator is

$$T_{\text{accel}} = J\frac{d\omega_{\text{r}}}{dt} = T_{\text{mech}} - T_{\text{e}} - T_{\text{loss}} \tag{6.12}$$

When acceleration torque $T_{\text{accel}} = 0$, the speed is constant and the generator is in the steady state. J is the combined moment of inertia of the turbine, shaft, and generator. When electrical load is added to the generator output terminals, the electrical torque T_{e} will increase, acceleration torque T_{accel} will become negative, and the generator shaft speed will decrease. In practical machines, the shaft speed is sensed by a governor. The governor senses this decrease in speed, and increases the flow to the turbine—steam for a steam turbine, water in the case of a hydro turbine, etc. A simplified diagram of a steam turbine generator shaft and governor is shown in figure 6.8.

In the steady state, the governor controller has droop R. In percent, R is the drop in frequency relative to system nominal frequency when the generator load goes from no load to rated load. Percent droops in the range of 3%–6% are most common. In a 60 Hz system, a 5% droop is 0.05 * 60 Hz = 3 Hz.

The droop constant S_R is defined as

$$S_R = \frac{\text{rated generator load}}{\text{rated frequency} * (\%R/100)} \tag{6.13}$$

The units of S_R are MW Hz^{-1} or kW Hz^{-1}, depending on the units used for the generator rating.

The generator controller set point is the no load speed command for the unit, f_{g}^*. The steady state generator g power output is then

$$P_{\text{g}} = S_{\text{Rg}}(f_{\text{g}}^* - f_{\text{system}}) \quad 0 < P_{\text{g}} < P_{\text{rated}} \tag{6.14}$$

This equation is valid over the permissible range of generator power output. As is the case with the voltage regulator, there are limiters and dynamic conditioners that do not impact the steady state response of the governor control. Equation (6.14) is shown in block diagram form in figure 6.9.

Speed droop is commonly applied to engine controls to reduce controller effort and improve stability. A controller that works to maintain the set frequency can overreact and lead to the unit 'hunting' for the correct operating speed/frequency. In some cases, a slower controller is used to move the engine generator back to set speed once the engine has responded to the change in load. This type of constant speed control is referred to as isochronous operation.

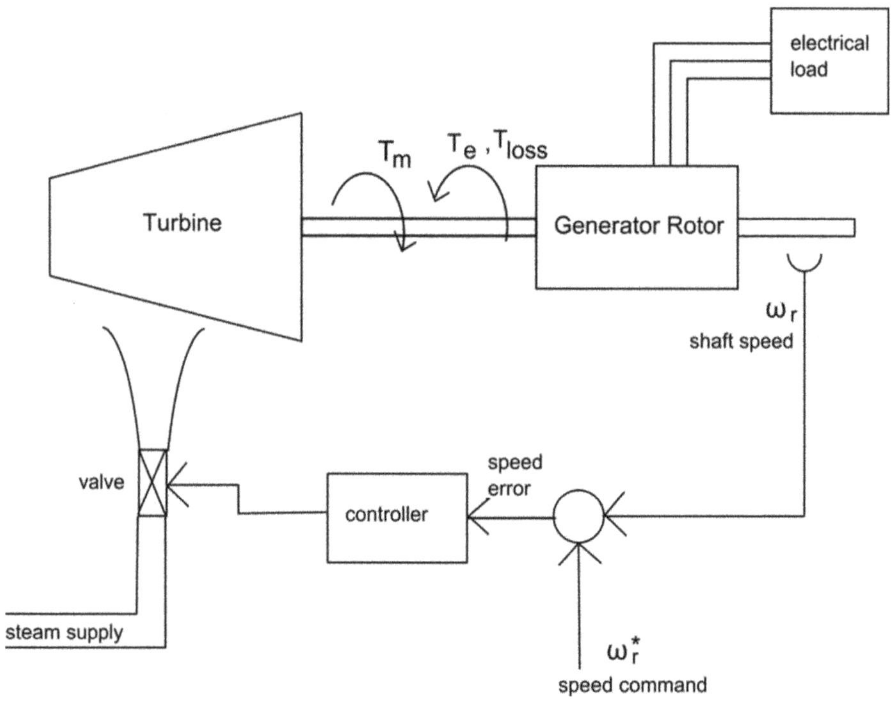

Figure 6.8. Conceptual diagram of steam turbine generator with speed governor.

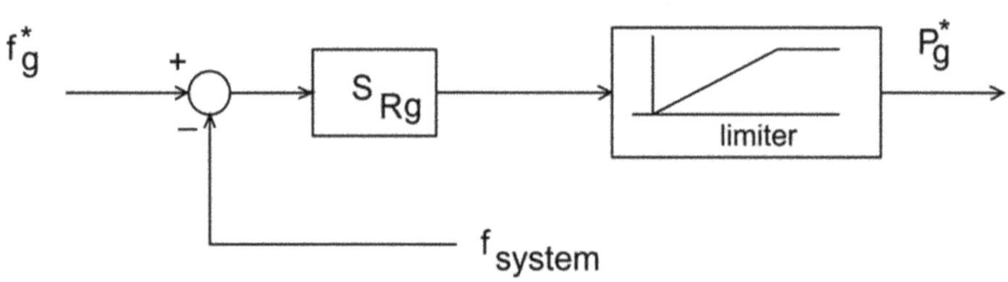

Figure 6.9. Block diagram of speed droop controller for generator g.

Example 6.5. The engine generator system of examples 6.2–6.4 is being operated with a speed governor that has a 5% speed droop characteristic. With the original $\overline{Z}_{wye} = 14\Omega$ load, the generator is operating at 1800 rpm. The generator real power rating is 2.5 MW and has four poles.

Find the frequency command for this operating point.

The system operating frequency is

$$f_{system} = 1800\,\text{rpm} \cdot \frac{2 \cdot 60}{4} = 60\,\text{Hz}$$

From equation (6.13), the droop constant is

$$S_R = \frac{2.5\,\text{MW}}{60\,\text{Hz} * (5/100)} = 0.833\,\text{MW}\,\text{Hz}^{-1}$$

From equation (6.14),

$$f_g^* = f_{system} + \frac{P_g}{SR_g} = 60.0 + \frac{1.234}{0.8333} = 61.48\,\text{Hz}$$

A second load is added in parallel to the first, resulting in a new load impedance of $\overline{Z}_{wye} = 7\,\Omega$. What is the new shaft speed of the engine/generator?

From example 6.4, the generator output power has increased to 2.046 MW. The governor equation will now be

$$2.046 = 0.833(61.48 - f_{system})$$

From this,

$$f_{system} = 59.02\,\text{Hz}$$

The shaft speed will then be

$$\text{rpm} = f_{system} \cdot \frac{2 \cdot 60}{\text{poles}} = 1770\,\text{rpm}.$$

6.4 Grid connected operation

In bulk power systems, numerous synchronous generators are connected in parallel, all working together to supply the load. When analysing the operation of a single generator, the system can be modeled as a constant voltage source connected to a bus through an impedance. This ideal voltage source is generally referred to as an infinite bus. An infinite bus is essentially the (three phase) Thevenin voltage behind the Thevenin equivalent impedance (also three phase) of all of the other generators and loads of the power system.

The elements in a power system are often represented by one line diagrams. The one line diagram for a generator feeding into an infinite bus through a transformer and transmission line is shown in figure 6.10. This diagram conveys much of the same information as a full three phase schematic circuit, but is much simpler. Figure 6.11 shows the per phase equivalent circuit that represents the three phase system shown in the one line diagram of figure 6.10. The impedance $R_{sys} + jX_{sys}$ in this figure would be the sum of the transformer impedance, the line impedance, and the infinite bus impedance. The impedances and the infinite bus voltage would need to be referred to the generator bus to get this equivalent circuit.

In order for the system to operate properly, each generator must have appropriate governor and excitation controls. The droop controllers outlined in the previous section provide this function and work independently at each generator. The system level controllers work by adjusting the voltage and power commands to these local controllers.

Example 6.6. The synchronous machine of example 6.2 is connected to the bulk power system, as shown in figures 6.10 and 6.11. The bulk power system can be represented by an infinite bus behind a per phase impedance of $\overline{Z}_{sys} = 0.04 + j\,0.32\,\Omega$. The infinite bus voltage is 4160 V/2400 V wye.

(a) Draw the per phase equivalent circuit of the system.

The equivalent circuit is shown below. Note that several variables are still to be determined.

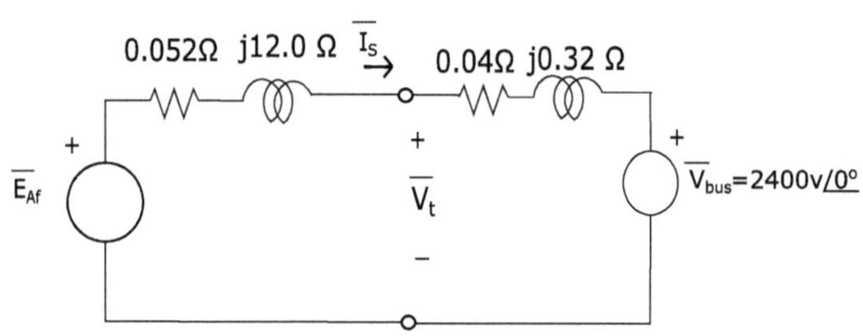

(b) The generator is delivering 3 MVA at 0.92 power factor lagging, as measured at the infinite bus. Determine the stator current $\overline{I}_s$, and the voltages $\overline{E}_{AF}$ and $\overline{V}_t$.

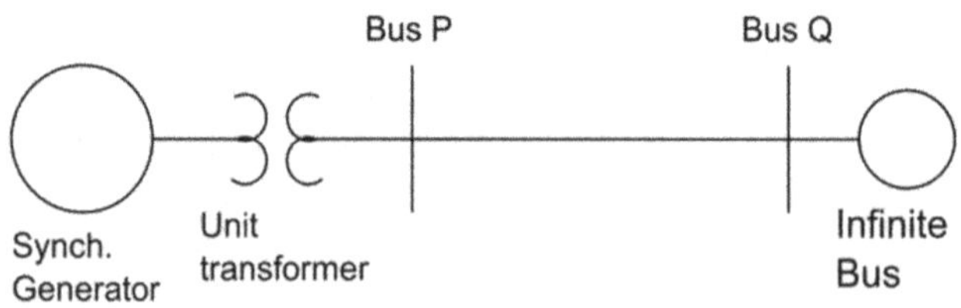

Figure 6.10. One line diagram of synchronous generator feeding into a large system through a transformer and transmission line.

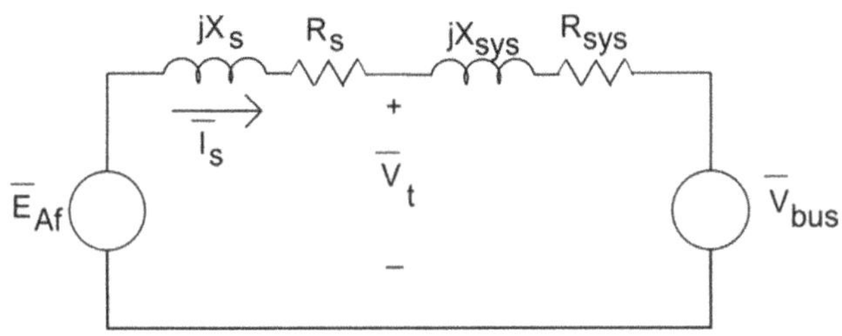

Figure 6.11. Per phase equivalent circuit for figure 6.10. The system impedance $R_{sys} + jX_{sys}$ is the equivalent impedance of the transformer, transmission line and system, all referred to the generator side of the unit transformer. Similarly, V_{bus} is the system voltage as seen from the generator side of the transformer.

The voltage is known at the infinite bus, so the magnitude of the stator current is

$$\bar{I}_s = \frac{S_{3\phi}}{3V_{LN}} = \frac{S_{3\phi}}{\sqrt{3}\,V_{LL}} = \frac{3.0 \times 10^6\ \text{VA}}{3 \cdot 2400\ \text{V}} = 417\ \text{A}$$

The power factor of this flow is 0.92 lagging. Therefore, this current will lag the infinite bus voltage by

$$\cos^{-1} 0.92 = 23.1^\circ$$

The stator current is then

$$\bar{I}_s = 417\underline{/-23.1^\circ}$$

The generator terminal voltage is then

$$\bar{V}_t = 2400\ \text{V}\underline{/0^\circ} + (0.04 + j0.32) \cdot 417\underline{/-23.1^\circ} = 2470\underline{/2.7^\circ}$$

The generator internal voltage is

$$\bar{E}_{AF} = 2470\text{V}\underline{/2.7^\circ} + (0.052 + j12) \cdot 417\underline{/-23.1^\circ} = 6476\ \text{V}\underline{/46.6^\circ}$$

Clearly, the impact of the generator synchronous impedance is greater than that of the power system in this case. However, it should be remembered that the generator internal voltage E_{AF} does not actually appear at the machine terminals—it is a representation of the voltage induced by the magnetic flux created by the field current. This field flux is counteracted by the flux created by the stator current.

However, in cases where the load is suddenly lost, the stator current drops significantly, and the terminal voltage can rise dramatically. This must be quickly counteracted by a reduction in the field current. The excitation control system must be able to do this effectively to avoid damage.

Excitation control: When the generator is feeding into a strong bus, the bus voltage will have a significant impact on the generator terminal voltage V_t. The excitation control will still adjust the value of the internal generator voltage E_{Af}, but the primary result of changing the value of E_{Af} will be to adjust the reactive power flow from the generator. If current lags voltage, the generator is supplying VARs to the load. This is called overexcited operation, and has relatively high values of excitation voltage E_{Af}. With lower values of excitation voltage E_{Af}, VARs are supplied to the generator by the system. These VARs increase the excitation level of the generator, replacing the reduced excitation supplied by the field winding. This is called underexcited operation, and has leading power factors, as the generator is absorbing VARs from the load. In the lagging power factor case, the field fully excites the machine, and excess excitation sends VARs to the system. At the point where the field excitation exactly matches the machine requirement, there is no transfer of VARs between stator and system, and the machine operates at unity power factor. The phasor diagrams of figure 6.5 apply to this case as well as the standalone case. However, in the standalone case, the power factor is set by the load impedance, while in this case the power factor and reactive power flow are primarily set by the generator internal voltage.

This relationship between generator field current and armature current is shown in figure 6.12. This figure shows a family of curves. These curves are referred to as generator V curves. Each curve represents a fixed real power output from the generator. The horizontal axis is field current. Remember that generator internal voltage E_{Af} is proportional to field current at constant saturation levels. At low field currents, the generator is underexcited, and the unit is operating a leading power factor with high phase current. The generator is drawing VARs from the system to increase its excitation level. As field current is increases, the stator current goes down, until the unit is operating at unity power factor. At this point, the field is providing all of the excitation for the machine. As field current continues to increase, the field excitation increases, and the generator starts delivering VARs to the system. The stator phase current starts to increase again. The generator is now operating at lagging power factor. It is overexcited and delivers the excess excitation to the

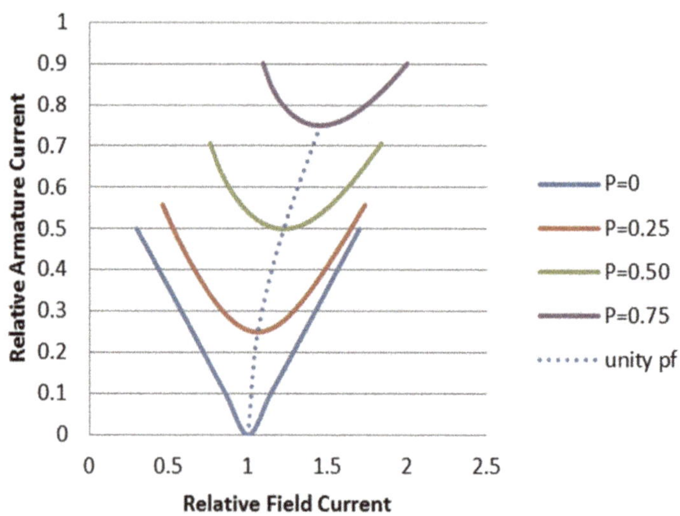

Figure 6.12. Synchronous generator V curve, showing the effect of field current on armature (stator) current.

system through VAR flow out of the machine. The $P = 0$ case is interesting—this is an application of synchronous machines known as a synchronous condenser. Synchronous condensors are strictly VAR sources—they operate without any mechanical load or driver, and provide or absorb reactive power as needed to maintain system voltages or adjust line flows.

In grid operation, generator terminal voltage and generator VAR output are closely coupled. Individual generators must guard against excessive VAR flow, whether into or out of the machine. However, as a group, the generators must maintain the system voltage and also must supply the reactive power needs of the system and load. In practice, many reactive power sources are located on the grid, and the generators only need to make up the remaining needs.

The droop control of the field controller creates the ability to share VARs between the many generators connected to the grid. If each generator were to attempt to control its terminal voltage to an exact value, the generators would be constantly fighting to exert their authority on the system, and individual generators would quickly be operating at maximum VAR levels, some producing the VARs and others absorbing them.

Maximum power transfer and the power-angle curve: The previous section shows that excitation current controls the generator reactive power flow in a grid connected synchronous generator. The real power flow from the generator is set by the torque supplied by the mechanical driver of the system. This mechanical driver is often a turbine, but also can be a piston engine or other energy source. The mechanical power flow into the generator is regulated by a governor. The governor regulates the shaft speed (or equivalently generator frequency). The governor acts to increase the turbine power flow when the speed drops. Through this action, it will maintain the generator speed within an acceptable range as the generator is loaded and unloaded.

For the one line diagram shown in figure 6.10, consider the case where there is no resistance in the impedance between the generator internal voltage and the infinite bus. The line current is found from the equation

$$E_{\mathrm{Af}} \underline{/\delta} = V_{\mathrm{bus}} \underline{/0^\circ} + jX_{\mathrm{T}} I_{\mathrm{A}} \underline{/\phi} \tag{6.15}$$

Here, X_{T} is the total inductive reactance of the system, the sum of the generator synchronous reactance and the Thevenin equivalent reactance of the system, representing the transformer as well as any line and system impedance, as seen from the generator terminals. Also for the moment, assume that the generator voltage regulator is turned off, and the internal voltage E_{Af} is constant.

Equation (6.15) can be separated into its real and imaginary parts,

$$\begin{aligned} E_{\mathrm{Af}} \cos\delta &= V_{\mathrm{bus}} - X_{\mathrm{T}} I_{\mathrm{A}} \sin\phi \\ E_{\mathrm{Af}} \sin\delta &= X_{\mathrm{T}} I_{\mathrm{A}} \cos\phi \end{aligned} \tag{6.16}$$

The real power coming out of the generator terminals is

$$P_{\mathrm{gen}} = 3E_{\mathrm{Af}} I_{\mathrm{A}} \cos\phi \tag{6.17}$$

By solving the second part of equation (6.16) for current, and substituting the result into the power equation (6.17),

$$P_{gen} = \frac{3E_{Af}V_{bus}}{X_T}\sin\delta \tag{6.18}$$

Recall that the generator power is set by the turbine and its governor. When E_{Af} and V_{bus} are constant, equation (6.18) says that the electrical system will accept the power output from the generator by adjusting the operating angle δ between the generator internal voltage and the infinite bus. A plot of equation (6.18) is shown in figure 6.13. The maximum power transfer between generator and system is

$$P_{MAX} = \frac{3E_{Af}V_{bus}}{X_T} \tag{6.19}$$

Figure 6.13 shows the steady state operating point for an initial generator power P_0, with the system operating at angle δ_0 If the turbine controls increase the generator power to P_1, the unit will respond by increasing the voltage angle to a value of δ_1. The movement of the generator from δ_0 to δ_1 will be governed by the torque equation (5.50)—the rotor will accelerate and the angle will increase. In response to this increased angle, the electric power out of the generator will increase, eventually matching the new mechanical input power.

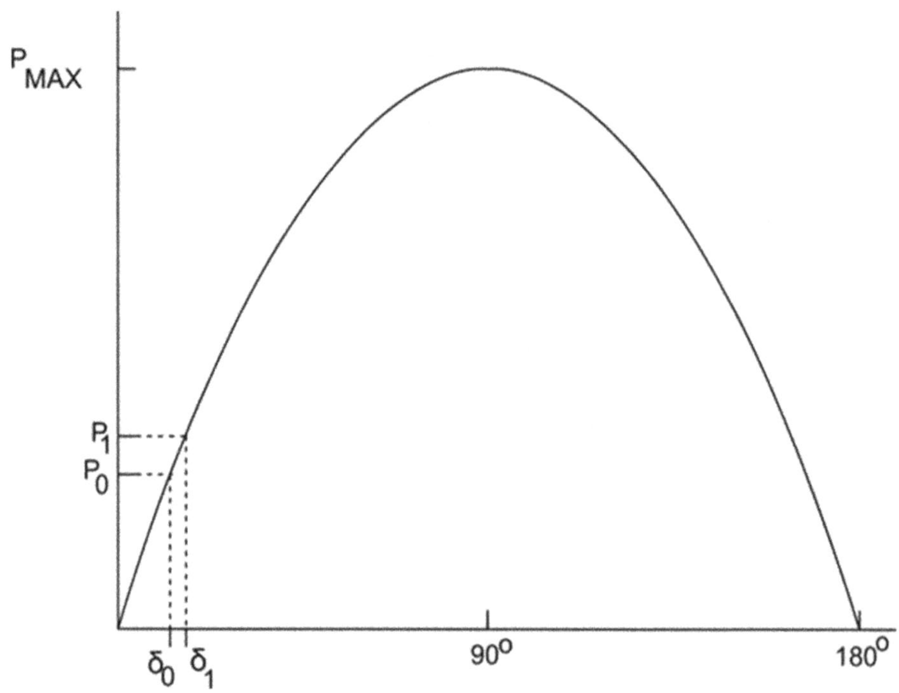

Figure 6.13. Generator power/angle curve showing operating angle change when the generator power increases from P_0 to P_1.

In this manner, the generator power can be increased up to a point where P_{gen} approaches P_{MAX}. If P_{gen} were to become greater than P_{MAX}, the system cannot absorb the power delivered by the generator.

In practice, the voltage regulator and governor controls will have some impact on this power versus angle relationship. With a fast voltage regulator, the machine terminal voltage could be considered constant, and an alternate steady state power-angle curve could be developed based on terminal voltage,

$$P_{MAX} = \frac{3V_t V_{bus}}{X_{sys}} \tag{6.20}$$

In this case, the generator's voltage regulator is continuously adjusting E_{Af} in order to keep terminal voltage V_t constant. While this relationship is strictly true only in the steady state, it is approximately true when the system power flows change slowly.

Example 6.7. In practice, the generator power is set by the governor control, and the field current is set by the excitation current. It is not unusual to need to solve the circuit when the real power flow, the excitation voltage, and the infinite bus voltage are known.

The generator is supplying 2.25 MW to the infinite bus. The infinite bus line to neutral voltage is 2400 V and the generator internal line to neutral voltage is 5200 V. Find the stator current, generator terminal voltage, and the VARs flowing at the generator terminals. Generator and power system resistance can be neglected.

From equation (6.18),

$$P_{gen} = 2.25 \times 10^6 = \frac{3E_{AF}V_t}{X_T}\sin\delta = \frac{3 \cdot 5200 \cdot 2400}{12.32}\sin\delta$$

From this equation,

$$\delta = \sin^{-1}\left(\frac{2.25 \times 10^6}{3.039 \times 10^6}\right) = 47.8^\circ$$

The stator current can then be found as

$$\overline{I}_s = I_s \underline{/\phi_s} = \frac{\overline{E}_{AF} - \overline{V}_t}{jX_T} = \frac{5200\,\underline{/47.8^\circ} - 2400\,\underline{/0^\circ}}{j12.32} = 324\text{ A}\,\underline{/-15.9^\circ}$$

The terminal voltage can be found from either the infinite bus voltage or the internal voltage. From the infinite bus voltage,

$$\overline{V}_t = V_t\underline{/\delta_t} = \overline{V}_{bus} + jX_{sys} \cdot \overline{I}_s = 2400\,\underline{/0^\circ} + j0.32 \cdot 324\,\underline{/-15.9^\circ}$$

The resulting line to neutral terminal voltage is

$$\overline{V}_t = 2430\text{ V}\,\underline{/2.4^\circ}$$

From equation (6.7), the VAR flow at the generator terminals is

$$Q_{gen} = 3V_t I_s \sin(\delta_t - \phi_s) = 0.74\,\text{MVAR}$$

Note that in this is a scalar equation, where V_t and I_s are the magnitudes of terminal voltage and phase current. The angles δ_t and ϕ_s are the respective angles of these phasors.

Generator dispatch in multi-machine systems: When considering an entire power grid, the governor controls on each unit must coordinate to provide power sharing among generators as the grid load rises and falls. The operating equation for the governor of generator k is

$$P_k = S_{Rk}(f^*_{nlk} - f_{sys}) \tag{6.21}$$

The term S_{Rk} is the droop constant of unit k, from equation (6.13). It is the steady state power/frequency relationship of unit k, set by the governor. It is stated in terms of kilowatts/hertz (kW Hz^{-1}) or megawatts/hertz (MW Hz^{-1}). Each unit will individually have their own no load frequency set point, f^*_{nlk}. The system frequency f_{sys} will be the same for each unit, as all generators in the system must operate at the same frequency (apart from the small short term adjustments to move up and down the power-angle curve).

A grid with numerous generators is shown in figure 6.14. With system resistance neglected, the total power generated by the N generators must match to total system demand, which is the sum of all of the loads on the system,

$$\sum_{k=1}^{N} P_{gk} = P_{demand} \tag{6.22}$$

Equation (6.21) can be substituted into equation (6.22), with the result

$$\sum_{k=1}^{N} S_{Rk} f_{nlk} - f_{sys} \sum_{k=1}^{N} S_{Rk} = P_{demand} \tag{6.23}$$

The system frequency for a given demand is then

$$f_{sys} = \frac{\sum_{k=1}^{N} S_{Rk} f_{nlk} - P_{demand}}{\sum_{k=1}^{N} S_{Rk}} \tag{6.24}$$

Once system frequency is known, the individual generator power P_{gk} can be found from equation (6.21).

In practice, this system of local governor actions works well in allowing a large group of generators to supply the always changing demand of large power grids. A balancing authority (such as New York Independent System Operator) adjusts the generation mix to minimize cost while maintaining reliability. They do this by sending periodic changes in the frequency command signal to the generators. These changes can come as often as every minute or so. During normal operation, system

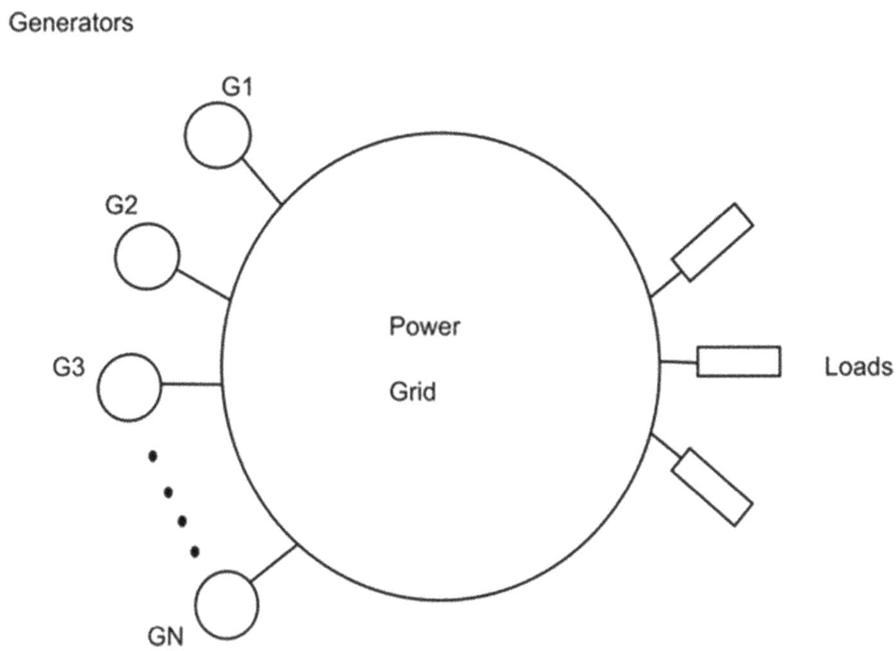

Figure 6.14. Conceptual diagram of *N* synchronous generators feeding multiple loads through the power grid.

operators will send these changes to some of the generators, and others will maintain steady power output through a fixed command frequency command.

The governors on synchronous machine turbine generator or engine generator systems are typically set for a 3%–6% speed droop, where speed droop can be determined from either the generator shaft speed n or operating frequency f,

$$\%\text{speed droop} = \frac{n_{\text{no load}} - n_{\text{full load}}}{n_{\text{nominal}}} * 100\% = \frac{f_{\text{no load}} - f_{\text{full load}}}{f_{\text{nominal}}} * 100\% \tag{6.25}$$

A speed droop of 5% on a nominal 60 Hz system would give a drop in frequency of 3 Hz when a generator goes from no load to full load. A 5% speed droop is the most common setting on grid connected synchronous generators in North America.

In this context, full load is generally considered to be the real power P at maximum allowable steady state loading. This maximum load can be set by the generator, the turbine, or other system parameter.

The droop constant for generator k, SR_k, can be related to percent droop.

$$SR_k = \frac{\text{generator k full load power}}{\text{speed droop in hertz}} = \frac{\text{generator k full load power}}{f_{\text{nominal}}\left(\frac{\%\text{speed droop}}{100}\right)} \tag{6.26}$$

When the regulation constant $\%R$ is the same for all of the generators in a system, they will share real power in proportion to the machine rating. During system operation with increasing load, the generators will increase their generation based on their respective droop constants. The balancing authority will then redispatch the generation on order keep generators, transformers and lines within their power ratings, and also to minimize the cost of generation.

Example 6.8. A two generator 60 Hz power system is shown below. With Load 1 connected, the system is observed to operate at 60.4 Hz, and the generator powers are shown in the diagram. Determine the system frequency and the two generator power flows when Load 2 is switched on. Assume that the system is lossless.

From equation (6.13),

$$S_{R1} = \frac{50\ \text{MW}}{60\ \text{Hz}(5/100)} = 16.67\ \text{MW Hz}^{-1}$$

$$S_{R2} = \frac{35\ \text{MW}}{60\ \text{Hz}(4/100)} = 14.58\ \text{MW Hz}^{-1}$$

With just Load 1 connected, the generator set point frequencies can be found from equation (6.14),

$$P_{G1} = S_{R1}(f_{G1}^{*} - f_{\text{system}})$$

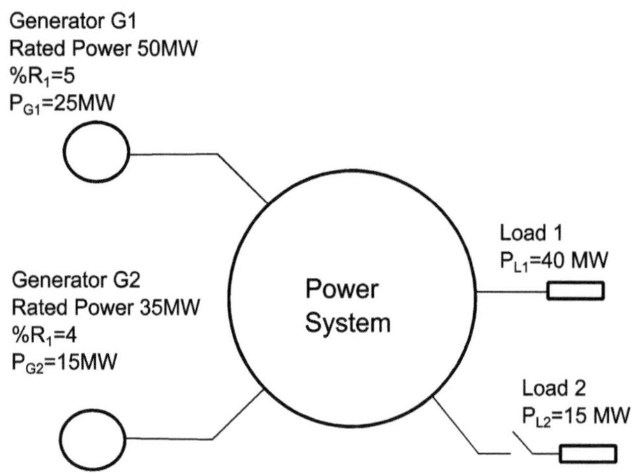

For Generator 1, $f_{G1}^{*} = 60.4 + \frac{25}{16.67} = 61.90\ \text{Hz}$

For Generator 2, $f_{G2}^{*} = 60.4 + \frac{15}{14.58} = 61.43\ \text{Hz}$

These set point frequencies remain constant in the face of changing load power.

When Load 2 is switched on, the system load goes up from 40 to 55 MW. Initially, the generator power remains constant, and as a result the system frequency will drop. This will bring the governor controls into action. The governors will increase their respective generator output powers until the generator power again equals the load power. The system will eventually settle at a new steady state operating point at a new system frequency.

From equation (6.21),

$$P_{G1} + P_{G2} = P_{\text{demand}} = 55\ \text{MW}$$

Substituting equation (6.14) into this equation,

$$S_{R1}(f_{R1}^{*} - f_{sys2}) + S_{R2}(f_{R2}^{*} - f_{sys2}) = P_{demand}$$

Here, f_{sys2} is the new system operating frequency. Substituting values,

$$16.667(61.90\text{ Hz} - f_{sys2}) + 14.58(61.43\text{ Hz} - f_{sys2}) = 55\text{ MW}$$

Solving for the new system frequency,

$$f_{sys2} = 59.92\text{ Hz}$$

Knowing the system frequency, the individual generator powers can be found,

$$P_{G1} = 33.0\text{ MW}$$

$$P_{G2} = 22.0\text{ MW}$$

As a check, the sum of these two generator powers does match the new load power.

Synchronizing: The starting of a large turbine generator unit is an interesting process, and can take minutes or hours or even a few days, depending on the energy source. From the generator point of view, once the energy is available valves can be opened and the turbine will begin to rotate the shaft. In a large steam turbine unit, this is done in a controlled and relatively slow manner. At some point as speed increases, the field circuit is energized, and the machine starts to build up field flux. The tasks of the voltage regulator and speed governor are now to bring the voltage magnitude and frequency to near system levels. When the voltage magnitude is within bounds, a synchronizing relay is enabled. This relay monitors the voltage across the open circuit breaker. This is illustrated in figure 6.15. Since the generator frequency and the system frequency are not identical, the voltage across the open

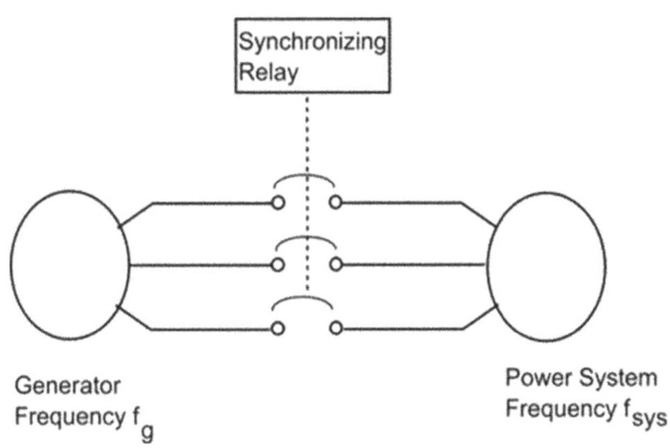

Figure 6.15. Synchronous generator with synchronizing relay to connect the generator to the power system when the frequencies match.

circuit breaker will vary with the slip frequency, which is the difference between the generator and system frequencies. When the slip frequency is sufficiently low (ideally a fraction of a hertz), the circuit breaker can be closed when the instantaneous voltage magnitude across the breaker is sufficiently small. Large circuit breakers can take several milliseconds to close, so modern synchronizing relays measure the rate of change of the slip frequency, and initiate a close so that the voltage is minimum when the breaker contacts close. Once the circuit breaker contacts close, the electrical coupling between system and generator locks the generator shaft speed to the corresponding synchronous speed. If the generator is not synchronized properly, large current and torque transients can occur. In extreme cases, these could be sufficient to damage the generator.

Questions

1. FNET is a wide area frequency measurement network operated by the Power Information Technology Laboratory at University of Tennessee. Its website is http://fnetpublic.utk.edu/index.html. It shows real time frequency and phase angle information for the power grid.
   a. Bring up the US frequency gradient map. In reading this, note that the US and Canada have four distinct interconnection areas, the Eastern Interconnect, the Western Interconnect, Texas, and Quebec. The nominal frequency of each area is 60 Hz, but each area can have a slightly different operating frequency. Observe the map for several minutes. Record the maximum and minimum frequencies you observe on the eastern interconnect.
   b. At a given instance, record the frequency reported for each of the four operating areas. Which area has the highest frequency at this instant? Which has the lowest? What is the difference in hertz between highest and lowest frequencies that you observe?
2. The electric utility industry is a regulated industry. The Federal Energy Regulatory Commission (FERC) provides market oversight for interstate electricity markets. As a part of this, the US is divided into several regional electric power markets. Many other parts of the world have similar regional power markets.
   a. Identify a regional electricity market.
   b. Describe the region covered by this market.
   c. Briefly describe the generation and transmission assets of this market.

Problems

1. A three phase, four pole 60 Hz synchronous generator is rated at 800 kVA, 480 V/277 V wye. The generator synchronous reactance $X_s = 0.8$ ohms and the stator resistance is $R_s = 0.004$ ohms. The generator internal voltage magnitude is $E_{Af} = \omega_e K_e I_F$, where $K_e = 0.035$. The field resistance is $R_F = 8\ \Omega$. The generator is operating at 60 Hz.

a. The generator is initially supplying power to a wye connected load with impedance of $\overline{Z}_{\text{wye}} = 1.2 + j0.3\ \Omega$. The terminal voltage is $\overline{V}_{\text{t}} = 270\text{V}$. Determine the stator current from the per phase diagram.
b. Determine the internal phasor voltage $\overline{E}_{\text{Af}}$.
c. Find the field current and the field voltage that will provide this value of E_{Af}.
d. A new load is added in parallel to the original load. The impedance of this load is $\overline{Z}_{\text{wye2}} = 1.8 + j0.5\ \Omega$. The value of E_{Af} remains the same. What is the new terminal voltage?
e. What is the voltage droop in this case?

2. A voltage regulator is added to a synchronous generator for automatic field control. The voltage regulator is shown in figure 6.6. The generator is operating at 60 Hz, and the generator field resistance is $R_{\text{F}} = 4\ \Omega$. $K_{\text{e}} = 0.022$ for the machine.
   a. The generator is originally loaded with a terminal voltage magnitude of $V_{\text{t}} = 275$ V. In the controller, $V_{\text{t}}^{*} = 280$ V and $V_{\text{f0}} = 150$ V. The controller has a constant gain of 2.5. From this, find the field voltage, the field current, and E_{Af}.
   b. What is the ratio of $V_{\text{t}}/E_{\text{Af}}$? This can be thought of as the steady state 'voltage gain' of the generator, from a control point of view.
   c. The load is changed, and after the field controller has adjusted the field voltage, the terminal voltage magnitude V_{t} has dropped to 265 V. The controller inputs and gain remain the same. Determine the new value of E_{Af}, and the resulting generator voltage gain.
3. A synchronous generator is feeding an impedance load. The generator real power rating is 1400 kW. It is operating with a governor, and the governor has a frequency droop setting of $\%R = 4\%$.
   a. Find the droop constant S_R for this machine.
   b. Initially, the generator output is $P_{\text{g}} = P_{\text{g}}^{*} = 400$ kW. It has a command frequency of $f_{\text{g}}^{*} = 60.5$ Hz. From figure 6.9, find the generator operating frequency f_{system}.
   c. The generator output is increased to 900 kW. Find the new operating frequency.
   d. This is a four pole machine. Find the change in generator shaft speed in rpm.
4. The per phase diagram of a synchronous generator connected to an infinite bus is shown below. The generator is a wye connected 66 MVA, 60 Hz 38 pole machine driven by a hydraulic turbine. The generator is operating with a terminal voltage of 13.2 kV L-L (7.62 kV L-N). The generator is operating with rated output volt–amps at an 0.85 lagging power factor. The system frequency is 60.0 Hz.
   a. Find the phasor values of internal voltage E_{a} and phase current I_{A}.
   b. Find the generator shaft speed and developed torque.

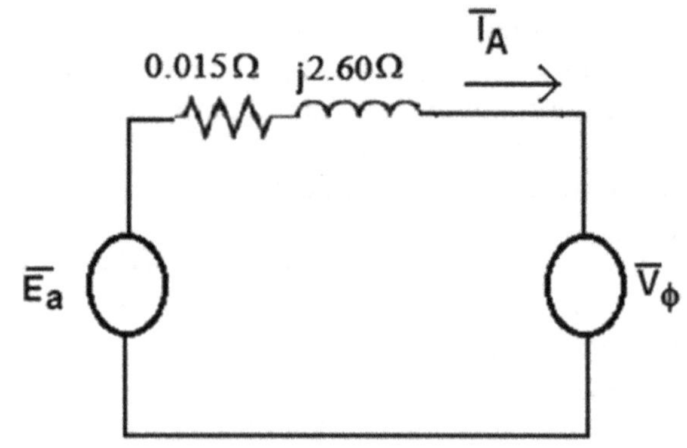

5. A power grid is fed by two synchronous generators. Assume that the power grid has no losses. The generator governor control settings are:

Generator 1	$\%R_1 = 2$	$P_{rated} = 600$ MW	$f_1{}^* = 60.1$
Generator 2	$\%R_2 = 6$	$P_{rated} = 250$ MW	$f_2{}^* = 60.5$

The system frequency is $f_{sys} = 59.5$ Hz.
 a. Calculate the two generator output powers, and the total load on the system.
 b. The load increases by 75 MW. Calculate the drop in system frequency and the resulting output of the two generators.
 c. Is either generator overloaded after this addition of this new load?
6. A three generator system is serving 300 MW of load. The nominal system frequency is 60 Hz.
 - Generator 1 has a rating of 200 MW and has a regulation constant of $\%R_1 = 5\%$. Generator 1's operating frequency set point is $f_1{}^* = 60.05$ Hz.
 - Generator 2 has a rating of 100 MW and a regulation constant of $\%R_2 = 5\%$. Generator 2's operating frequency set point is $f_2{}^* = 60.10$ Hz.
 - Generator 3 has a rating of 150 MW and has a regulation constant of $\%R_1 = 5\%$. Generator 3's operating frequency set point is $f_3{}^* = 60.15$.
   a. Determine the system operating frequency.
   b. Find the power delivered by each of the three generators.
7. A three phase 60 Hz synchronous motor has four poles. The motor rated voltage is 460 V/266 V wye. The motor rated stator current is 175 A. The motor internal voltage is $E_{Af} = 28I_f$, where I_f is the motor field current. The per phase equivalent circuit is shown below.

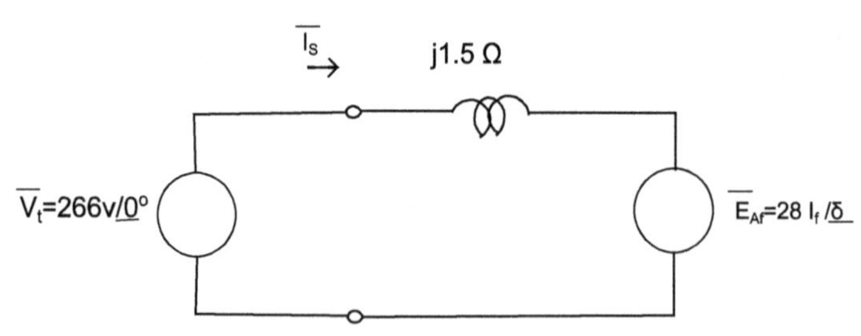

Develop the V curve for this motor, when the real component of I_s is 150 A, plotting stator current as a function of field current. Include at least five data points in the V curve. To solve this, vary the out of phase component of I_s over a range of positive and negative values, while keeping the magnitude of I_s less than 175 A. Then solve for E_{Af}, and then I_f. To get the full V curve, include both points where I_s =175 A (with both lagging and leading current), and the point where I_s = 150 A and is in phase with V_t. It is convenient to use MATLAB or other analytical software to do the repetitive calculations. Use a plotting routine from MATLAB, Excel or other software to present your results.

8. A three phase, four pole 60 Hz synchronous generator is rated at 650 MVA and 22.5 kV/13 kV wye. The generator synchronous reactance X_s = 0.90 ohms. The generator is feeding into a system that has system equivalent reactance X_{sys} = 0.15 ohms and infinite bus voltage of 13.0 kV line to neutral.
   a. The generator internal voltage E_{af} = 19.5 kV line to neutral. Determine P_{max} for this operating point.
   b. The generator is delivering 550 MW to the infinite bus. Determine the angle δ of the generator internal voltage.
   c. Determine the generator terminal voltage V_t.
   d. Determine the VAR flow at the generator terminal for this operating point.
   e. The generator power is increased to 600 MW. Determine the new value of the power-angle δ.

IOP Publishing

Electromechanical Machinery Theory and Performance

Thomas Ortmeyer

Chapter 7

Induction machines

7.1 Overview

Induction machines hold a wide variety of roles in today's society. Squirrel cage induction motors offer low cost and high reliability, and are applied in sizes from fractional horsepower to over 10 000 horsepower. Induction motors directly connected to the power grid provide reasonable starting torque and low speed droop. Induction motors fed by rectifier/inverters are competitive with other variable speed technologies. Induction machines are currently being used in wind turbine technologies, in a variety of forms, using both direct connection and interfacing to the grid with power electronics. They are also one of the machine options used in electric vehicles.

Induction machines have either a wound rotor or a squirrel cage rotor. The squirrel cage rotor consists of rotor bars in slots along the length of the rotor. The bars are connected by end rings at both ends, shorting the entire set of bars. In a three phase squirrel cage induction motor, stator currents create a rotating flux wave, just as in a synchronous machine. The stator flux wave induces currents in the squirrel cage bars on the rotor. These rotor currents in turn create a flux wave of their own. Torque is created to align the stator and rotor flux vectors created by the stator and rotor currents.

In wound rotor machines, current is induced in the rotor windings in the same manner as the currents induced into a squirrel cage. These windings are connected to the stationary world by slip rings. These rotor windings are generally three phase, and three slip rings are generally used. For a wye wound rotor, the neutral point is generally not brought out.

The induction machine model in this chapter is developed from the point of view of a wound rotor machine. This model will then be revised to represent squirrel cage machines as well.

doi:10.1088/978-0-7503-1662-0ch7

7.2 Theory

Standard induction machines have a set of three phase stator windings similar to synchronous machines. Consider for the moment a *2 pole machine*, as shown in figure 7.1. When excited by balanced three phase currents, the stator windings create a rotating flux wave of constant magnitude and rotating at a speed proportional to the electrical frequency f_1 of the stator currents,

$$\omega_1 = 2\pi f_1 \tag{7.1}$$

On the induction machine rotor, consider another set of three phase coils. These are similar to the stator coils—balanced three phase currents of frequency f_2 hertz flowing in these coils will create a rotating flux wave that rotates at a speed of ω_2 rad/sec, where $\omega_2 = 2\pi f_2$. This flux rotates with respect to a fixed position on the rotor. Notice that subscript '1' is used for the stator windings, and '2' for the rotor set of windings.

The rotor itself, however, is moving at speed $\omega_r = \frac{d\theta_r}{dt}$ rad/sec. From a stationary reference, the rotor flux wave is rotating at a speed of

$$\frac{\text{poles}}{2}\omega_r + \omega_2 \tag{7.2}$$

This is illustrated in figure 7.2, for the two pole case. Equation (7.2) generalizes this to machines with larger numbers of pole pairs.

As was learned with synchronous machines, the stator and rotor rotating flux waves produce a torque proportional to the sine of the angle between them. A constant

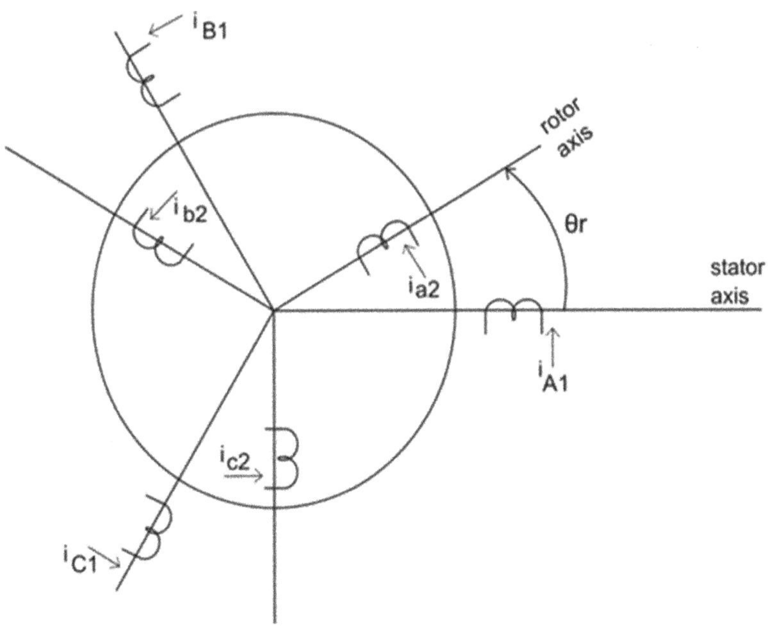

Figure 7.1. Diagram of induction machine showing the three stator phases and three rotor phases.

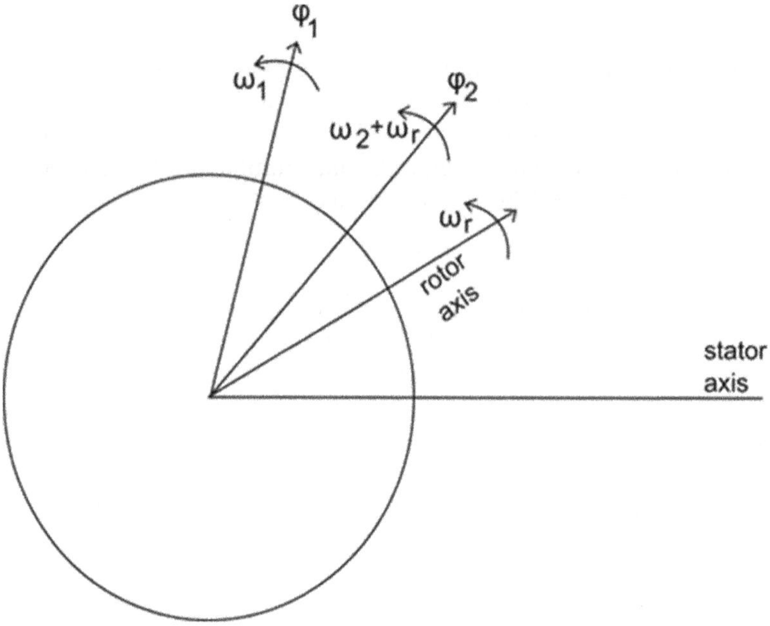

Figure 7.2. Rotating flux waves set up by stator (1) currents and rotor (2) currents.

torque is therefore developed when the angle between these flux waves is constant. This can only happen when the two flux waves are rotating at the same speed,

$$\omega_1 = \frac{\text{poles}}{2}\omega_r + \omega_2 \tag{7.3}$$

A wound rotor induction machine can actually be used as a frequency changer, with the frequency relationship given by equation (7.3). Mechanical shaft torque would be required to provide power balance for the two sets of windings. With present day power electronics, this is not a particularly practical application of the technology. The wound rotor induction motor also becomes a synchronous machine when $\omega_1 = \frac{\text{poles}}{2}\omega_r$, and the rotor frequency $\omega_2 = 0$. In this case, DC current externally supplied to the rotor winding will excite the machine.

7.3 Stator and rotor rotating flux waves

Consider for the moment that the rotor windings are open circuited. The three phase stator winding is the same as the three phase stator winding in the synchronous machine. As is the case with synchronous machine stator windings, balanced three phase currents in the induction motor stator winding will produce a rotating air gap flux wave. From section 5.4, if the A phase stator current is $\sqrt{2}\,I_1\cos(\omega_1 t+\phi_e)$, this flux wave will be

$$\Phi_1 = \frac{3K}{\sqrt{2}} I_1 \cos(\omega_1 t - \gamma_e + \phi_e) \tag{7.4}$$

This equation is valid for balanced sinusoidal three phase currents in the stator with constant magnitude $\frac{3K}{\sqrt{2}}I_1$ and rotational speed ω_1. At any point in time t the location of the peak stator flux is at angle $\gamma_e = \omega_1 t + \phi_e$.

This rotating magnetic flux wave will produce flux linking each of the stator coils. When $\gamma_e = 0$, the rotating flux wave aligns with the stator A phase winding, and all of the flux Φ_1 links this winding. This flux linkage then varies with the cosine of the angle where the Φ_1 maximum lies. Therefore, the component of Φ_1 linking the A phase stator winding is

$$\Phi_{A1} = \frac{3K}{\sqrt{2}} I_1 \cos(\omega_1 t + \phi_e) \tag{7.5}$$

This flux induces a voltage on the A phase winding,

$$e_{A1} = N_1 \frac{d\Phi_{A1}}{dt} = \frac{3KN_1}{2} \frac{d\sqrt{2}I_1 \cos(\omega_1 t + \phi_e)}{dt} \tag{7.6}$$

The flux Φ_{A1} is directly proportional to the magnitude of the stator currents. The induction motor magnetizing inductance is then determined to be

$$L_m = \frac{3KN_1}{2} \tag{7.7}$$

The current I_1 that flows when the rotor is open circuited goes to magnetize the core. It is convenient to label this magnetizing current as I_{1m}.

This rotating flux wave aligns with the stator B phase winding when $\gamma_e = \frac{2\pi}{3}$ and the stator C phase winding at $\gamma_e = \frac{4\pi}{3}$. Therefore, the B and C phase induced voltages will be

$$\begin{aligned} e_{B1} &= N_1 \frac{d\Phi_{B1}}{dt} = \frac{3KN_1}{2} \frac{d\sqrt{2}I_1 \cos\left(\omega_1 t + \phi_e - \frac{2\pi}{3}\right)}{dt} \\ e_{C1} &= N_1 \frac{d\Phi_{C1}}{dt} = \frac{3KN_1}{2} \frac{d\sqrt{2}I_1 \cos\left(\omega_1 t + \phi_e - \frac{4\pi}{3}\right)}{dt} \end{aligned} \tag{7.8}$$

As is the case with transformers and synchronous machines, the stator winding also has both coil resistance and leakage inductance. These two elements create a voltage drop between the terminal voltage v_{A1} and the internal voltage e_{A1}. An equivalent circuit of the stator A phase winding is shown in figure 7.3. In this circuit, the magnetizing current is shown as the current flowing through the magnetizing inductance. In the phasor version of this circuit, this would correspond to a reactance of $X_m = \omega_1 L_m$

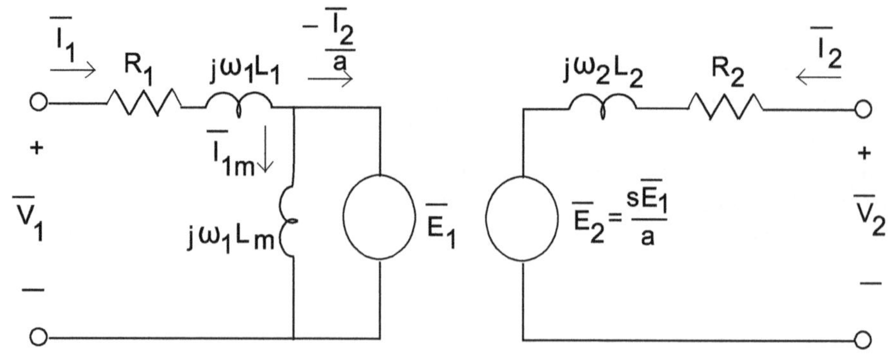

(a) Stator phasor diagram, frequency f_1 (b) Rotor phasor diagram, frequency f_2

Figure 7.3. Induction motor per phase equivalent circuit, with phasor quantities in the actual frequencies.

This same flux wave (equation (7.4)) generated by the stator current will link the rotor coils,

$$\Phi_1 = \frac{3K}{\sqrt{2}} I_1 \cos(\omega_1 t - \gamma_e + \phi_e) \tag{7.9}$$

With the rotor moving at speed ω_r, the rotor windings will see this rotating wave as moving at speed

$$\omega_2 = \omega_1 - \frac{\text{poles}}{2} \omega_r \tag{7.10}$$

This is shown in figure 7.2. If, for example, the stator frequency is 60 Hz, the Φ_1 wave will rotate 60 revolutions per second. If the rotor is rotating at 59 revolutions per second, then the windings on the rotor will see only 1 cycle of flux per second. This will induce 1 Hz voltages on the rotor. If rotor winding currents flow, these will create a rotating flux wave rotating at 1 Hz relative to the rotor and 60 Hz relative to the stator.

In general, the rotor electrical frequency equals the difference between the stator electric frequency and the rotor mechanical speed.

$$f_2 = f_1 - \frac{\text{poles}}{2} f_r \tag{7.11}$$

which is the same as equation (7.10), with units in hertz. This frequency f_2 is often called the slip frequency—with the flux wave rotating at f_1 hertz and the rotor shaft rotating at $\frac{\text{poles}}{2} f_r$ revolutions per second, the flux wave 'slips' by the rotor windings f_2 times per second.

If at some arbitrary time $t = 0$, the stator and rotor windings are aligned and $\gamma_e = 0$, then the full value of flux created by the stator current will link the rotor A phase winding. As time continues, the flux linking this winding will vary with the slip frequency,

$$\Phi_{A2} = \frac{3K}{\sqrt{2}} I_1 \cos(\omega_2 t + \phi_e) \tag{7.12}$$

The induced voltage in this winding will then be

$$e_{A2} = N_2 \frac{d\Phi_{A1}}{dt} = \frac{3KN_2}{2} \frac{d\sqrt{2}\, I_1 \cos(\omega_2 t + \phi_e)}{dt} \tag{7.13}$$

In this equation, N_2 is the number of turns in each of the three rotor windings. Comparing this equation with equation (7.6),

$$\begin{aligned} e_{A1} &= N_1 \frac{d\Phi_{A1}}{dt} = -\frac{3\omega_1 K N_1}{2} \sqrt{2}\, I_1 \sin(\omega_1 t + \phi_e) \\ e_{A2} &= N_2 \frac{d\Phi_{A1}}{dt} = -\frac{3\omega_2 K N_2}{2} \sqrt{2}\, I_1 \sin(\omega_2 t + \phi_e) \end{aligned} \tag{7.14}$$

As these two voltages are at different frequencies, there is no direct relationship between their instantaneous values. The relationship between their amplitudes, however, is

$$E_{A2} = \frac{N_2}{N_1} \frac{\omega_2}{\omega_1} E_{A1} \tag{7.15}$$

Remember, the frequency ω_2 is the frequency of the rotor voltages and currents. It is called either the rotor electrical frequency or the slip frequency. The term slip is defined as the ratio of the slip frequency to the stator frequency,

$$s = \frac{\omega_2}{\omega_1} \tag{7.16}$$

If the turns ratio of the machine is defined as for a transformer,

$$a = \frac{N_1}{N_2} \tag{7.17}$$

Then

$$E_{A2} = \frac{s}{a} E_{A1} \tag{7.18}$$

The rotor B and C phases will have the same magnitude as the A phase rotor voltage, and will be at lagging at angles of $\frac{2\pi}{3}$ and $\frac{4\pi}{3}$ respectively—these angles being the electrical angles at the slip frequency.

Each rotor winding will have a leakage inductance and winding resistance, the same as for the stator windings and for transformers. When the rotor winding terminals are connected to a balanced impedance, the rotor winding currents that flow will be balanced. The form of the A phase current will be

$$i_{A2} = \sqrt{2}\, I_2 \cos(\omega_2 t + \phi_2) \tag{7.19}$$

The reference direction for this current is assumed to be into the rotor winding. This balanced set of rotor winding currents will in turn create a rotating flux wave in the air gap of the machine. This flux wave will be

$$\Phi_2 = \frac{3K_2}{\sqrt{2}} I_2 \cos(\omega_2 t - \gamma_2 + \phi_2) \tag{7.20}$$

Equation (7.20) shows that Φ_2 rotates at speed ω_2 when measured from the *rotor* A phase magnetic axis. γ_2 is an arbitrary position measure from the rotor A phase magnetic axis. Since this axis is rotating at speed ω_r, and $\omega_1 = \omega_2 + \frac{\text{poles}}{2}\omega_r$, the rotor traveling flux wave will be

$$\Phi_2 = \frac{3K_2}{\sqrt{2}} I_2 \cos(\omega_1 t - \gamma_1 + \phi_2) \tag{7.21}$$

Φ_2 then has a rotational speed of ω_1 when measured from the A phase *stator* reference axis. It is important to note that under these conditions, this flux wave is traveling at the same speed as the stator flux wave Φ_1 and that there will be a constant angle between these two rotating flux waves in the steady state. This is illustrated in figure 7.2.

These conditions—stator and rotor flux waves having constant magnitudes and rotating at the same speed, is necessary for the induction machine to being able to generate a constant torque. The torque produced is proportional to the vector cross product of these two flux waves.

If current is allowed to flow in the rotor circuit, the magnetic circuit requires that it be matched by current in the stator—the same as occurs in transformers. As in a transformer, when the magnetizing inductance L_m is modeled separately, the magnetic path can be considered to be ideal and the MMF required to create the flux is vanishingly small. Ideally, the MMF relationship between stator and rotor currents is then

$$\mathcal{F} = 0 = N_1 i_1 + N_2 i_2 \tag{7.22}$$

Including the magnetizing inductance, this equation becomes

$$i_1 = L_m i_{1m} - \frac{N_2}{N_1} i_2 = L_m i_{1m} - \frac{i_2}{a} \tag{7.23}$$

This is the final link for the full induction motor equivalent circuit. This circuit is shown in phasor form in figure 7.3. This figure shows the steady state phasor version of this circuit.

This circuit is interesting in several ways. In particular, the stator and rotor phasor circuits are operating at different frequencies. The currents in the stator and rotor circuits are related by the turns ratio between rotor and stator circuits. The induced voltages are related by the turns ratio as well, but is also related by the ratio of the rotor and stator frequencies. This ratio is defined as the slip s. The currents, however, are related by the turns ratio only.

This diagram represents both wound rotor and squirrel cage induction machines. Wound rotor machines are being used today as doubly fed machines. In these machines, the stator winding is generally connected directly to the power grid. A variable voltage, variable frequency power electronic converter is connected to the rotor. These generators are one of several competing technologies in the wind turbine industry. The equivalent circuit shown in figure 7.3 is useful for doubly fed machine applications.

Squirrel cage induction motors are the workhorse machine for direct grid connected motor applications. They are also used in power electronically fed variable speed applications. The squirrel cage rotor uses rotor bars rather than coils. The bars are shorted at both ends of the rotors. While it does not have three discrete phase windings on the rotor, this rotor is modeled as a three phase rotor with the rotor terminals short circuited. In this machine, the stator currents induce a voltage on the rotor. This voltage drives current flow in the shorted rotor bars. This rotor current in turn creates a magnetic field that interacts with the stator flux wave to create torque.

In a squirrel cage induction motor, it is worthwhile to refer the rotor quantities to the stator side and stator frequency.

The rotor circuit equation from figure 7.3(b) is $\overline{E}_2 = \overline{V}_2 - (R_2 + j\omega_2 L_2)\overline{I}_2$. If this equation is divided by slip s, the result is

$$\frac{\overline{E}_2}{s} = \frac{\overline{V}_2}{s} - \left(\frac{R_2}{s} + j\frac{\omega_2}{s}L_2\right)\overline{I}_2 \tag{7.24}$$

Note that $\frac{\overline{E}_2}{s} = \frac{\overline{E}_1}{a}$ and $\frac{\omega_2}{s} = \omega_1$. Inserting these, equation (7.24) becomes

$$\frac{\overline{E}_1}{a} = \frac{\overline{V}_2}{s} - \left(\frac{R_2}{s} + j\omega_1 L_2\right)\overline{I}_2 \tag{7.25}$$

If the rotor current is referred to the stator by dividing it by the turns ratio a, the equation then becomes

$$\overline{E}_1 = \frac{a\overline{E}_2}{s} = \frac{\overline{E}_2'}{s} = \frac{a\overline{V}_2}{s} - a^2\left(\frac{R_2}{s} + j\omega_1 L_2\right)\frac{\overline{I}_2}{a} = \frac{\overline{V}_2'}{s} - \left(\frac{R_2'}{s} + j\omega_1 L_2'\right)\overline{I}_2' \tag{7.26}$$

Here, $R_2' = a^2 R_2$ and $L_2' = a^2 L_2$ are the rotor resistance and inductance referred to the stator. The rotor current $\overline{I}_2$ and rotor terminal voltage $\overline{V}_2$ are referred to the primary winding as was done with transformers,

$$\begin{aligned} \overline{E}'_2 &= a\overline{E}_2 \\ \overline{I}'_2 &= \frac{1}{a}\overline{I}_2 \end{aligned} \tag{7.27}$$

The revised rotor circuit is shown in figure 7.4. This circuit represents the relationship of equation (7.26).

With this change, the internal voltages in the stator and the rotor circuit are the same, and the current entering the rotor internal voltage is the same current that leaves the stator internal voltage. These two circuits can then be combined. The result is shown in figure 7.5. The primed quantities in this figure indicate quantities that have been referred to the stator. Not only does this figure involve quantities that are referred to the stator, it also appears to be a phasor circuit operating at stator frequency ω_1.

This equivalent circuit is valid for both wound rotor and squirrel cage induction machines. However, for squirrel cage induction machines, the rotor bars are shorted, so $V_2' = 0$.

Core loss: Induction machines experience both hysteresis and eddy current loss in their cores. The bulk of this loss occurs in the stator of the machine, as the stator flux wave varies at the input frequency f_1. The flux variation on the rotor iron core varies at slip frequency f_2. The slip frequency is generally much lower than the input frequency during normal operation.

To a good approximation, the hysteresis loss is proportional to frequency, and the eddy current loss is proportional to frequency squared. Both vary with the square of core flux level.

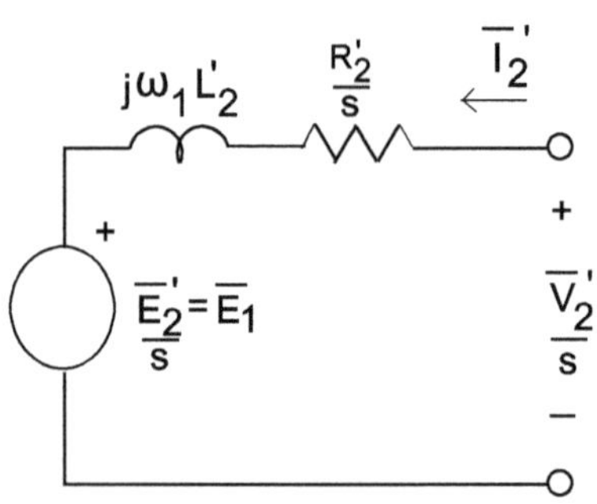

Figure 7.4. Revised rotor equivalent circuit referring it to stator quantities.

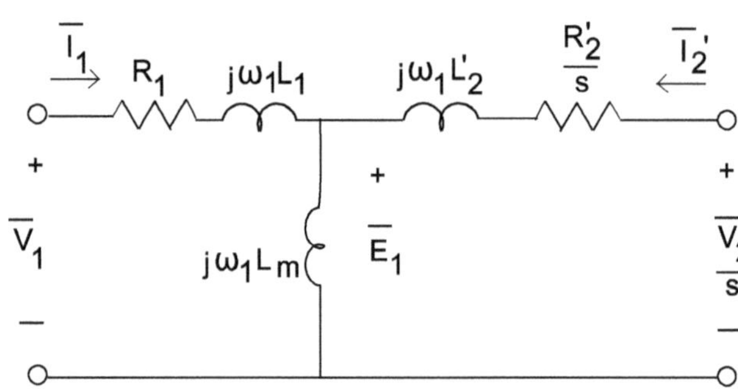

Figure 7.5. Revised induction machine per phase equivalent circuit (motor convention) with all quantities referred to the stator.

In induction motors, as in synchronous machines, it is common practice to include the core losses with the mechanical losses, and not model them in the equivalent circuit. This simplifies the analysis of the induction motor equivalent circuit, without significant loss of accuracy.

Induction machines that are directly connected to the power grid operate with near constant stator voltage and frequency, and it is generally acceptable to consider the core loss to be constant under these conditions. The next level of analysis would be to estimate the variation of core loss with supply voltage. The above approximations are reasonable when the supply voltage fluctuates within the ±5% tolerances typical of the power system. With larger changes in voltage and frequency, the induction motor core losses need to be adjusted accordingly, particularly in studies involving motor losses and efficiency.

Example 7.1. Doubly fed machines are essentially induction machines with wound rotors. Three phase voltages are applied to both the stator and the rotor. The frequencies of the stator and rotor voltages are generally different, and are related to the rotational speed by equation (7.11).

A three phase, two pole doubly fed machine is operating at 3240 rpm. The machine has a turns ratio of $a = 3$. The stator is fed by a balanced three phase 60 Hz source of 208 V line–line (120 V line to neutral). Use the per phase equivalent circuit of figure 7.3 to analyse this machine. Ignore the magnetizing reactance of the machine in this example.

(a) The machine is operating with the rotor terminals open circuited. What is the magnitude and frequency of the line to neutral rotor voltage V_2?

With $I_2 = 0$, and neglecting the magnetizing branch, $I_1 = 0$. Therefore $E_1 = V_1 = 120\text{V}$.

At 3200 rpm, the rotor rotates $\frac{3240}{60} = 54$ times per second. For this two pole machine, this can be considered the rotor rotational speed in electrical hertz.

From equation (7.11), the electrical frequency on the rotor circuit is

$$f_2 = f_1 - f_r = 60 - 54 = 6 \text{ Hz}.$$

The motor slip is then

$$s = \frac{6}{60} = 0.10$$

From figure 7.3, the internal rotor voltage is

$$\overline{E}_2 = \frac{s\overline{E}_1}{a} = \frac{0.1 \cdot 120 \text{ V}}{3} = 4 \text{ V}$$

Since $\overline{I}_2 = 0$, $\overline{V}_2 = \overline{E}_2 = 4\text{V}$.

(b) Also neglect the stator and rotor coil resistance and leakage inductance. If the stator current is 10 A in each phase, and is in phase with the stator voltage, calculate the rotor phase current, the three phase power entering both the stator and rotor, and the sum of these two powers. What happens to this resultant power?

Assume that the A phase stator voltage and stator current are at an angle of 0°. The three phase power entering the stator is

$$P_s = 3 \cdot 120\text{ V} \cdot 10\text{ A} \cdot \cos(0^\circ) = 3600\text{ W}$$

From figure 7.3, the rotor current is

$$\bar{I}_2 = -a\bar{I}_1 = 30\text{ A}\,\underline{/180^\circ}$$

The rotor winding power is then

$$P_r = 3 \cdot \bar{V}_2\bar{I}_2\cos(\delta_2 - \phi_2) = 3 \cdot 4\text{V} \cdot 30\text{A} \cdot \cos(0^\circ - 180^\circ) = -360\text{ W}$$

This is −360W ENTERING the rotor winding, or 360W LEAVING the rotor. The net electrical power *ENTERING* the machine is then

$$P_{\text{electrical}} = P_s + P_r = 3600 - 360 = 3240\text{ W}$$

As we have neglected any resistive losses in the machine, this power is not converted to losses. It therefore must be converted to mechanical power.

7.4 Torque and power

Recall that in the synchronous machine, the three phase power entering the internal emf was the electrical power that was converted to mechanical power. In the induction motor, electrical power can be converted from both the stator and rotor windings. From the per phase equivalent circuit of figure 7.3, the electric power entering the three phase stator winding has three components:

- Stator winding loss.
- Magnetizing reactive power.
- Converted power.

The stator side converted power is the power entering the internal voltage $\bar{E}_1$. This power is

$$P_{\text{conv1}} = 3E_1I_1\cos(\delta_{\text{E1}} - \phi_1) = 3E_1\left(-\frac{I_2}{a}\right)\cos(\delta_{\text{E1}} - \phi_2) \tag{7.28}$$

The second term is the power written in terms of the stator terminal voltage less the magnetizing current.

In the three phase rotor circuit, the power converted from electrical to mechanical is

$$P_{\text{conv2}} = 3E_2I_2\cos(\delta_{\text{E2}} - \phi_2) \tag{7.29}$$

In these equations, δ_{E1} and δ_{E2} are the angles of the internal stator and rotor voltages, which equal each other. Because $E'_2I'_2 = E_2I_2$,

$$P_{conv2} = 3E'_2I'_2\cos(\delta_{E2} - \phi_2) = (sE_1)\left(\frac{I_2}{a}\right)\cos(\delta_{E1} - \phi_2) = -sP_{conv1} \tag{7.30}$$

This equation states that the magnitude of the rotor converted power is the slip times the stator converted power, with rotor converted power flowing in the opposite direction of the stator converted power when slip is positive, and in the same direction when the slip is negative. For example, for a doubly fed generator with power leaving the stator, power must be supplied to the rotor for positive slips. For negative slips (rotor speed above synchronous speed), the power would be supplied by the rotor.

With the reference directions chosen in this chapter, both of these powers are positive going into the magnetic field, and work to increase the energy stored in this field. In the motor convention shown in figure 7.5, the mechanical power works to reduce the energy in the magnetic field. When these three converted powers balance, the energy in the magnetic field is constant, and the machine is in the steady state.

When the machine is operating in the steady state with a constant magnetic field energy, the balance of converted power is

$$P_{mechanical} = P_{conv1} + P_{conv2} = (1 - s)P_{conv1} \tag{7.31}$$

The relationship between these three powers is shown in figure 7.6, for the case where the stator power P_{conv1} is constant over the speed range. At slip $s = 1$, the rotor is at a standstill with no mechanical power output, and all the stator power (less losses) comes out of the rotor windings. As speed increases (slip decreases), the rotor power decreases, and the mechanical power increases. This continues until $s = 0$,

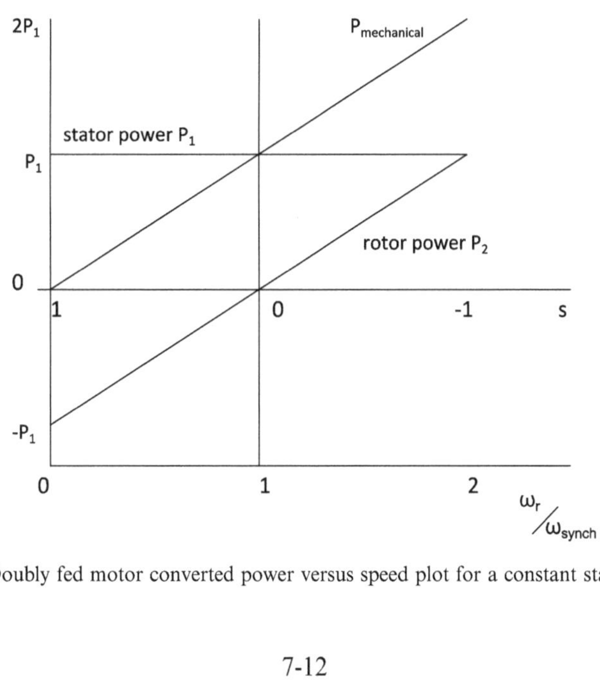

Figure 7.6. Doubly fed motor converted power versus speed plot for a constant stator power P_1.

when the machine is at synchronous speed and the rotor frequency is zero. At that point, the rotor goes from delivering power ($P_2 < 0$) to absorbing power ($P_2 > 0$). The mechanical power continues to increase with speed, up to the maximum speed of the machine.

From equation (7.31), the induction machine electrically developed torque is

$$T_e = \frac{1}{\omega_r}(1-s)P_{conv1} = \frac{2/\text{poles}}{\omega_1(1-s)}(1-s)P_{conv1} \quad (7.32)$$

Because

$$\begin{aligned} \omega_1 &= \frac{\text{poles}}{2}\frac{\omega_r}{(1-s)} \\ T_e &= \frac{\text{poles}}{2\omega_1}P_{conv1} \end{aligned} \quad (7.33)$$

So for this machine, torque T_e varies directly with stator power (when operating with a fixed stator frequency), and is independent of rotor shaft speed. This is a unique feature of the doubly fed machine.

A typical application is shown in figure 7.7. The doubly fed machine is acting as a motor, with the stator winding connected to a fixed frequency, fixed voltage power grid. The rotor circuit is connected to a variable speed variable voltage source—supplied by a pulse width modulated (PWM) electronic circuit fed by a DC source.

In figure 7.7, the loss term $T_{loss} = B\omega_r$ represents the friction and windage mechanical loss plus the magnetic core loss. The motor will operate in the

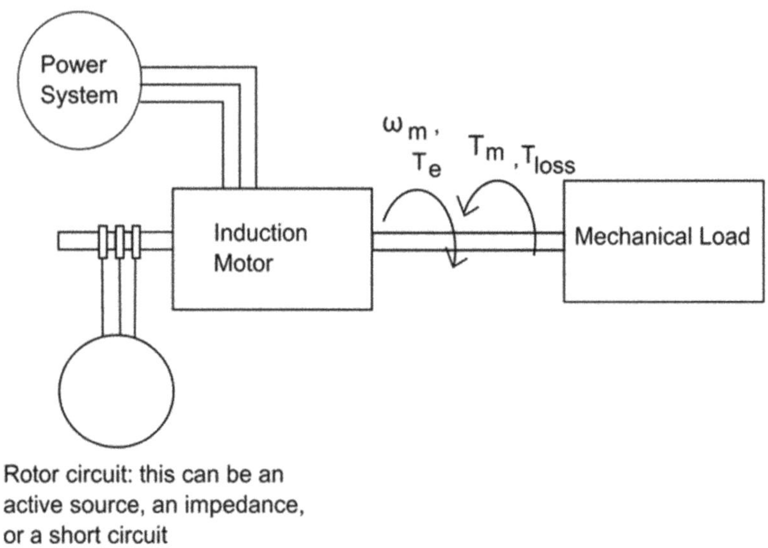

Figure 7.7. Induction motor connected to a mechanical load. Electrical power is positive entering the motor electrically, and leaving the motor mechanically.

steady state at constant speed when the torques balance so that there is no acceleration,

$$T_e = T_m + B\omega_r \tag{7.34}$$

During transients, this equation becomes

$$J\frac{d\omega_r}{dt} = T_e - T_m - B\omega_r \tag{7.35}$$

Here, ω_r is the shaft speed in mechanical radians per second. In terms of the stator frequency ω_1, slip and number of poles, it is

$$\omega_r = \frac{(1-s)\omega_1}{\text{poles}/2} \tag{7.36}$$

Example 7.2. Determine the rotational speed and developed torque of the doubly fed machine of example 7.1.

The machine in example 7.1 is producing 3240 W of mechanical power. This power is the product of rotor speed in mechanical radians per second and developed torque in newton-meters,

$$P_{\text{converted}} = \omega_r T_e$$

As this is a two pole machine, the rotor speed in radians per second is

$$\omega_r = 2\pi\frac{3240 \text{ rpm}}{60 \text{ s min}^{-1}} = 339.3 \text{ rad/sec}$$

The developed torque is then

$$T_e = \frac{P_{\text{converted}}}{\omega_r} = \frac{3240 \text{ W}}{339.3 \text{ rad/sec}} = 9.55\text{N m}$$

This is the internal torque developed in the machine, and is subject to rotational losses in the machine as stated in equation (7.34). The reference direction of this torque is shown in figure 7.6. With this convention, the mechanical power is positive when leaving the machine. In this example, 3240 W enters the machine's electrical terminals, and the same amount of power leaves the mechanical terminal of the machine, leaving no net loss. This 'ideal' power transformation cannot be realized in practice, of course, and electrical and mechanical losses mean that the conversion is not ideal.

7.5 Squirrel cage machines

The majority of induction motors have squirrel cage rotors. The squirrel cage rotor can be represented by three equivalent phase windings, with the winding terminals shorted. The equivalent circuit of this machine is obtained by shorting the rotor terminals of the equivalent circuit shown in figure 7.5. The resulting circuit is shown in figure 7.8. In this circuit, the machine inductances are shown as reactances. Also, for convenience, the reference direction for the rotor current has been reversed in this diagram,

$$\bar{I}'_r = -\bar{I}'_2 \tag{7.37}$$

From the figure 7.8 equivalent circuit, the power flow diagram for the squirrel cage induction machine operating in the motoring mode ($s > 0$) can be developed, and is shown in figure 7.9.

This figure shows that all of the power enters the motor through the stator. This input power less the stator copper loss is often referred to as the air gap power

$$P_{\text{airgap}} = P_1 - 3I_1^2R_1 = 3I_2'^2\frac{R_2'}{s}$$

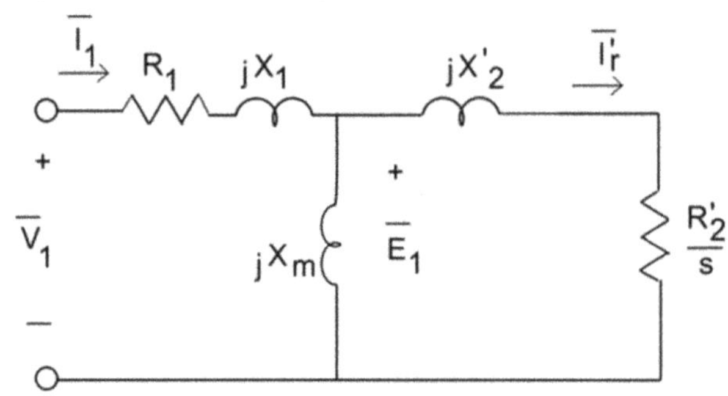

Figure 7.8. Per phase equivalent circuit for a squirrel cage induction motor. Note the new reference direction for rotor current.

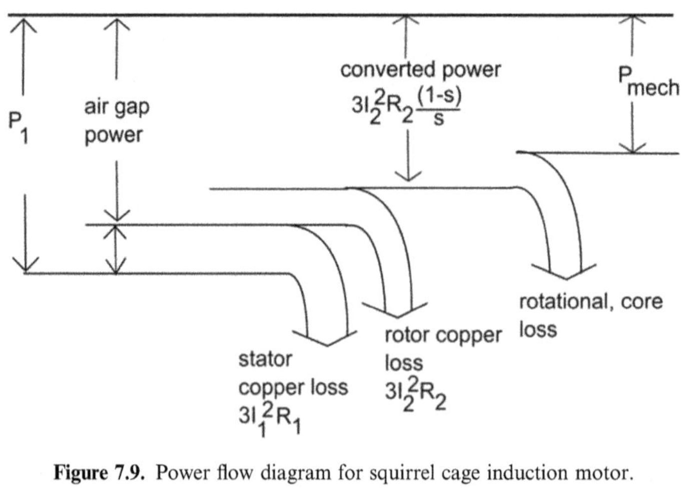

Figure 7.9. Power flow diagram for squirrel cage induction motor.

Note that in terms of power, the rotor power flows are the same (as they must be) whether using actual rotor quantities, or the rotor quantities as referred to the stator (primed quantities). The air gap power includes the power converted to mechanical power as well as the rotor copper loss. The rotor copper loss is

$$P_{\mathrm{RCL}} = 3I_2^2 R_2 = 3I_2'^2 R_2'$$

The converted power is then

$$P_{\mathrm{conv}} = P_{\mathrm{airgap}} - P_{\mathrm{RCL}} = 3I_2'^2 \frac{R_2'}{s} - 3I_2'^2 R_2' = 3I_2'^2 R_2' \frac{1-s}{s}$$

The electrically developed torque is then

$$T_{\mathrm{e}} = \frac{P_{\mathrm{conv}}}{\omega_{\mathrm{r}}} = \frac{3(\mathrm{poles})}{2\omega_{\mathrm{e}}} \frac{I_{\mathrm{r}}'^2 R_2'}{s} \tag{7.38}$$

Figure 7.9 suggests some other ways to calculate torque. Notice from figure 7.9 that the converted power is $\frac{1-s}{s}$ times the rotor copper loss, and together they equal the air gap power. Therefore,

$$P_{\mathrm{converted}} = (1 - s)P_{\mathrm{airgap}}$$

The developed torque is then

$$T_{\mathrm{e}} = \frac{P_{\mathrm{converted}}}{\omega_{\mathrm{r}}} = \frac{P_{\mathrm{airgap}}}{\omega_1\left(\frac{2}{\mathrm{poles}}\right)}$$

Note that the torque can be calculated from this formula without calculating the rotor current.

The output power is the mechanical converted power less the mechanical loss. In this model, the core loss is also included with the mechanical loss.

The motor slip expressed in terms of stator and rotor electrical frequency is given in equation (7.16). This can be rewritten to be in terms of the stator electrical frequency and the rotor shaft speed, with the shaft speed expressed in electrical radians per second $\left(\omega_{\mathrm{r}}' = \frac{\mathrm{poles}}{2}\omega_{\mathrm{r}}\right)$ or in hertz $\left(f_{\mathrm{r}}' = \frac{\mathrm{poles}}{2}f_{\mathrm{r}}\right)$.

$$s = \frac{\omega_2}{\omega_1} = \frac{\omega_1 - \omega_{\mathrm{r}}'}{\omega_1} = \frac{f_1 - f_{\mathrm{r}}'}{f_1} \tag{7.39}$$

With n being the shaft speed in revolutions per seconds (rpm), the equivalent electrical shaft speed in hertz is

$$f_{\mathrm{r}}' = \frac{\mathrm{poles}}{2} \frac{n}{60} \omega_{\mathrm{r}}' = 2\pi f_{\mathrm{r}}' \tag{7.40}$$

The synchronous speed of a machine in rpm is directly related to the input electrical frequency and the number of poles of the machine,

$$n_{\text{sync}} = \frac{60f_1}{\text{poles}/2} \tag{7.41}$$

The slip can then be conveniently written to as

$$s = \frac{n_{\text{sync}} - n}{n_{\text{sync}}} \tag{7.42}$$

Example 7.3. The equivalent circuit parameters of a 1755 rpm, 45 horsepower three phase four pole 60 Hz squirrel cage induction motor are shown in the table below.

R_1	X_1	X_M	R_2'	X_2'	Rated L–L voltage	Mechanical loss at rated speed
0.11 Ω	0.302 Ω	13.08 Ω	0.128 Ω	0.302 Ω	460 V	480 W

(a) Determine the rated line to neutral voltage

$$V_{\text{LN(rated)}} = \frac{460\ \text{V}}{\sqrt{3}} = 265.6\ \text{V}$$

(b) Determined the slip at rated speed.
From equation (7.39),

$$s_{\text{(rated)}} = \frac{1800\ \text{rpm} - 1755\ \text{rpm}}{1800\ \text{rpm}} = 0.025$$

(c) Find the stator and rotor currents at rated voltage and rated slip.
The currents are found from the impedance values from the above table and the per phase equivalent circuit of figure 7.8. With the voltage and slip known, there are several ways to solve this circuit. One approach is by impedance reduction. Using this method, the rotor impedance is

$$\overline{Z}_{\text{rotor}} = \frac{0.128}{0.025} + j0.302 = 5.12 + j0.302\Omega$$

The magnetizing impedance is

$$\overline{Z}_{\text{m}} = j13.08\Omega$$

The parallel combination of these two impedances is

$$\overline{Z}_{ag} = 4.267 + j1.928\Omega$$

The total motor impedance is then

$$\overline{Z}_{motor} = 0.11 + j0.302 + \overline{Z}_{ag} = 4.377 + j2.230\Omega$$

Assume that the stator voltage is at 0°. The stator current is then

$$\overline{I}_1 = \frac{\overline{V}_1}{\overline{Z}_{motor}} = \frac{265.6\text{V} \underline{/0^\circ}}{4.377 + j2.230\Omega} = 54.07\text{ A} \underline{/-27.1^\circ}$$

The rotor current can then be found from the current divider,

$$\overline{I'}_r = \overline{I}_1 \frac{\overline{Z}_m}{\overline{Z}_m + \overline{Z}_{rotor}} = 49.36\text{ A} \underline{/-6.1^\circ}$$

(d) Find the developed torque of the motor.
From equation (7.38),

$$T_e = \frac{3 \cdot 4}{2 \cdot 377} \frac{49.36^2 \cdot 0.128}{0.025} = 198.5\text{ N m}$$

(e) Find the motor efficiency at this operating point.
The motor input power is

$$P_{in} = 3 \cdot V_1 I_1 \cos(\delta_1 - \phi_1) = 38.38\text{ kW}$$

The rotor speed in mechanical radians per second is

$$\omega_r = \frac{2\pi}{60} 1755\text{rpm} = 183.8\text{ rad/sec}$$

The converted power is

$$P_{converted} = \omega_r T_e = 183.8\text{ rad/sec} \cdot 132.7\text{ N m} = 36.48\text{ kW}$$

The output power is

$$P_{out} = P_{converted} - P_{mech\ loss} = 36.48\text{ kW} - 0.48\text{ kW} = 36.00\text{ kW}$$

The motor efficiency is then

$$\eta = \frac{P_{out}}{P_{in}} = \frac{36.00\text{ kW}}{38.38\text{ kW}} = 93.8\%$$

(f) Find the input volt-amps, VARs, and power factor.
The input volt-amps are

$$S_{in} = 3 \cdot V_1 \cdot I_1 = 3 \cdot 265.6\text{ V} \cdot 54.07\text{ A} = 43.08\text{ kVA}$$

The input power factor is then

$$pf = \frac{P_{\text{in}}}{S_{\text{in}}} = \frac{38.38\ \text{kW}}{43.08\ \text{kVA}} = 0.876\ \textit{or}\ 87.6\%$$

The input VARs are

$$Q_{\text{in}} = \sqrt{S_{\text{in}}^2 - P_{\text{in}}^2} = 19.57\ \text{kVAR}$$

Note that this calculation for VARs does not give the sign. As the induction motor is inductive and inductors consume VARs, the value of Q_{in} is positive, signifying VARs entering the motor.

Example 7.4. Calculate the starting torque of the motor of example 7.3 at rated terminal voltage.

The starting torque is calculated at rotor speed $\omega_r = 0$, which gives a slip of $s = 1$. At this slip,

$$\overline{Z}_{\text{rotor}} = \frac{0.128}{1} + j0.302 = 0.128 + j0.302\Omega$$

The parallel combination of the rotor impedance and the magnetizing reactance is

$$\overline{Z}_{\text{ag}} = 0.1223 + j0.2964\Omega$$

The motor impedance is then

$$\overline{Z}_{\text{motor}} = 0.11 + j0.302 + \overline{Z}_{\text{ag}} = 0.2323 + j0.5984\Omega$$

The stator current is

$$\overline{I}_1 = \frac{\overline{V}_1}{\overline{Z}_{\text{motor}}} = \frac{265.6\text{V}\,\underline{/0^\circ}}{0.2323 + j0.5984\Omega} = 413.8\text{A}\,\underline{/-69.0^\circ}$$

The rotor current is then

$$\overline{I}_2 = \overline{I}_1 \frac{\overline{Z}_{\text{m}}}{\overline{Z}_{\text{m}} + \overline{Z}_{\text{rotor}}} = 404.4\text{A}\,\underline{/-68.5^\circ}$$

The torque is then

$$T_e = \frac{3 \cdot 4}{2 \cdot 377} \frac{404.4^2 \cdot 0.128}{1.0} = 333.2\text{N m}$$

(a) What is the ratio of the starting current to the running current?

The ratio of the starting stator current to running stator current is

$$\frac{413.8\text{A}}{54.07\text{A}} = 7.65$$

Note that the starting current is also at a much lower power factor than the running current. This is referred to as line start of the induction motor, and the large current draw can cause voltage sags when the power source has some impedance.

Example 7.5. The induction motor of the previous two examples is being fed from a source with open circuit voltage of 480 V/277 V. The source has an impedance of $\overline{Z}_{\text{thev}} = 0.10 + j0.40\Omega$ per phase.

(a) Find the stator current and the starting torque of the motor with this source.

From figure 7.11, the total impedance of the system is the sum of the source impedance and the motor impedance,

$$\overline{Z}_{\text{total}} = \overline{Z}_{\text{thev}} + \overline{Z}_{\text{motor}} = 0.4048 \ + \ j1.0009\Omega$$

The stator current is then

$$\overline{I}_1 = \frac{\overline{V}_{\text{th}}}{\overline{Z}_{\text{total}}} = \frac{277.0\text{V}\,\underline{/0^\circ}}{0.4048 + j1.0009\Omega} = 256.6\text{A}\,\underline{/-68.2^\circ}$$

The starting torque is then

$$T_e = 228.14 \text{ N m}$$

The starting torque is therefore reduced from 333.2 N m to 226.1 N m due to the source impedance. The stator current is reduced from 413.8 A to 256.6 A. The ratio of starting to running current is reduced to

$$\frac{256.6\text{A}}{54.07} = 4.74$$

Note that both the 'ideal' source of 265.6 V with no impedance and the more typical source voltage of 277 V behind and impedance of 0.1 + j0.4 ohms gives the same voltage and torque at the operating point of example 7.3. The ratio of starting to running current of just under 5. In practice, this ratio is typically in the range of 5–7. This example also shows that starting torque must be evaluated with the source impedance considered. If the source voltage is too low or the source impedance too high, the motor may not be able to accelerate the load, and motor damage can result.

7.6 Induction motor operation

The induction motor torque speed curve is shown in figure 7.10. This is found from the per phase equivalent circuit of figure 7.8, with a constant source voltage $\overline{V}_1$. The curve starts at zero speed and goes to synchronous speed for motor operation. Synchronous speed is defined as the motor shaft speed if the motor was a synchronous motor rather than an induction motor. In rpm, the synchronous speed is

$$n_{\text{sync}} = \frac{120 f_1}{\text{poles}} \tag{7.43}$$

The slip goes from 1 at standstill to 0 at synchronous speed. Both of these scales are shown on the horizontal axis. At a standstill, the slip of 1 means that the rotor electrical frequency f_2 equals the stator electrical frequency f_1. Low speed

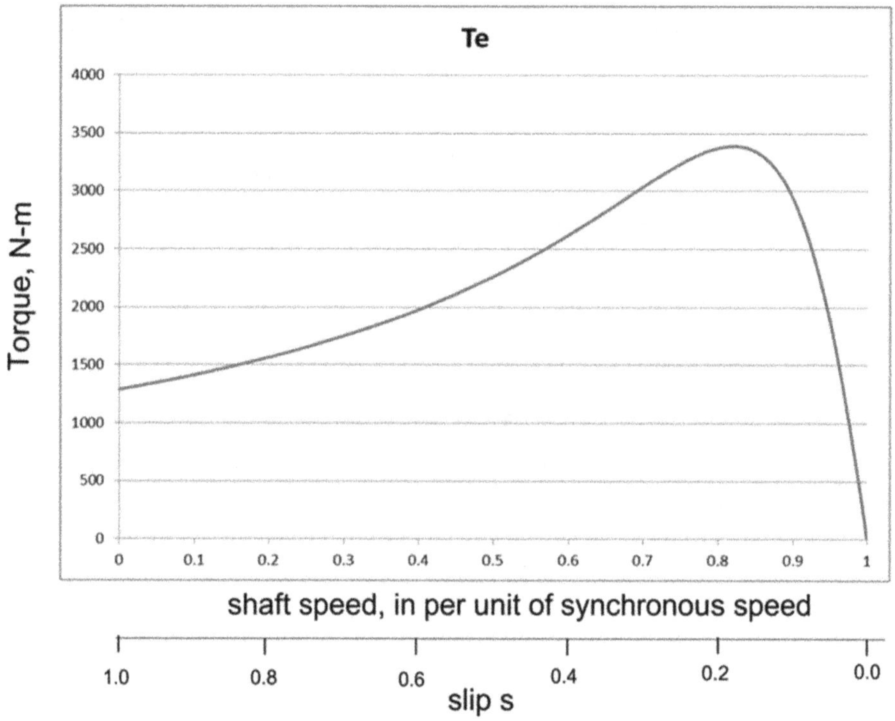

Figure 7.10. Induction motor torque-speed curve.

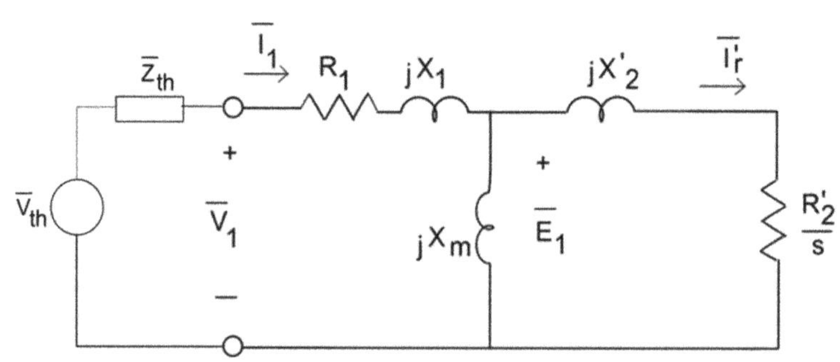

Figure 7.11. Induction motor per phase equivalent circuit with source impedance.

operation draws high currents at low power factors, and long term operation in this region is not sustainable or stable. Motors operate in this region during starting as they accelerate from standstill to operating speed. As speed increases, the motor reaches peak torque. At higher speeds, the motor enters its operating range. In this region, the motor operates with reasonable currents and power factor. At some point in this range, the motor enters a region where the motor can operate continuously in the steady state. At higher speeds yet, the torque decreases until it reaches zero torque at synchronous speed. For most motors, the motor operating slip is small, and the motor operates within a few percent of the motor's synchronous speed.

Table 7.1 shows the parameters of the motor whose torque–speed curve is shown in figure 7.10.

The loss torque consists of core loss and friction and windage losses. Figure 7.12 shows the electrical torque developed by the motor and the mechanical load torque T_{m}.

Motor loads: Induction motors are called on the drive many different loads. These can be divided into just a few general categories, however. These include:

Square law torque: Most fans and pumps have a torque speed characteristic of

$$T_{\mathrm{m}} = T_{\mathrm{m}o} + T_{\mathrm{m}2}\omega_{\mathrm{r}}^{2} \tag{7.44}$$

Aerodynamic loads also have torques that vary with the square of speed.

Constant torque: Elevators, lifts, conveyors, and similar equipment require a torque that is nearly constant as speed changes.

Variable torque: Some loads change with time and/or conditions. For example, a conveyor belt will operate with a constant torque over a range of speeds. When a box is dropped on the belt, it will move to a new value of constant torque. Another example is a vehicle traveling down a road and coming to a hill. The torque will increase as it climbs the hill. Still another example is a fan with variable outlet vanes, as the vanes close to restrict flow. Punch presses have an impact load—they require a large surge of torque for a short period of time. A piston compressor will have torque that fluctuates rapidly with the loading/unloading of each cylinder.

Table 7.1. Motor and system parameters for the figure 7.10 induction motor torque speed curve.

Parameter	Value	Ratings	
R_1	0.021 ohms	Source voltage	467 V L-L270 V L-n
X_1	0.072 ohms	Output power	250 horsepower
X_{m}	3.60 ohms	Speed	1753 rpm
R_2'	0.026 ohms	Poles	4
X_2'	0.072 ohms	Power factor	92.1%
B	28.5 watt-sec/radian	Efficiency	92.6%
		Frequency	60 Hz

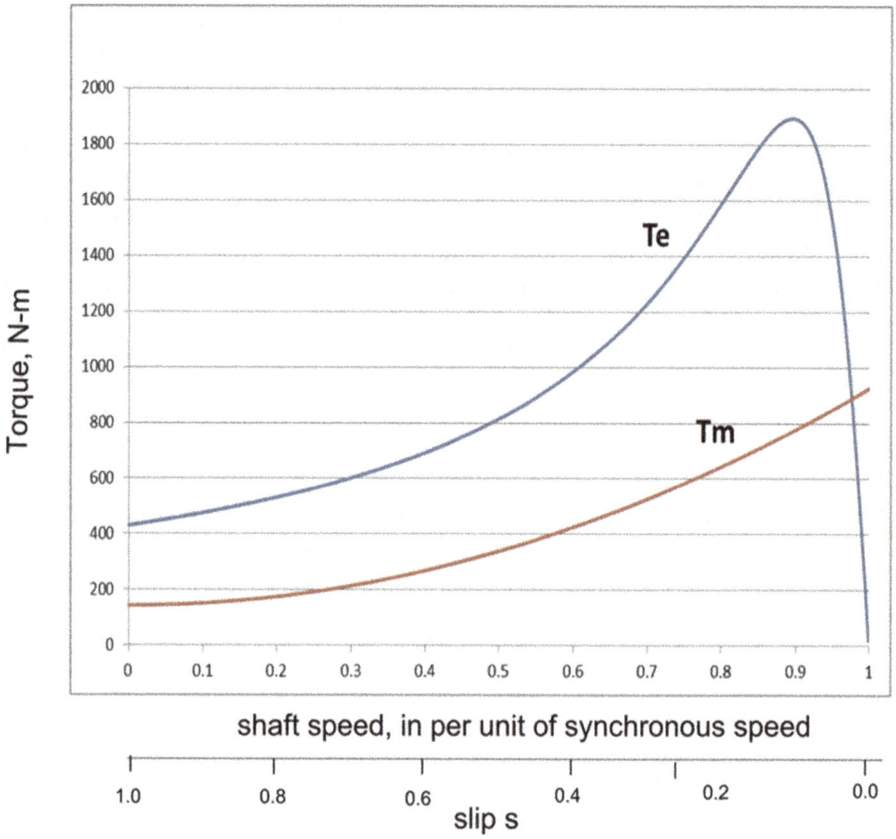

Figure 7.12. Table 7.1 induction motor with source impedance and load torque.

For successful operation, the induction motor must have sufficient starting torque and the capability of running continuously without overheating for the required load torque.

Motor ratings and capability: Individual induction motors come with a set of ratings that describe their characteristics and capability. The major ratings are shown in table 7.1. The machine ratings will also include additional information on insulation, and frame size, as well as thermal and operating limits.

Source impedance: In practice, the power system that supplies the induction motor is not ideal. The power system can be represented by its Thevenin equivalent voltage and impedance in the per phase equivalent circuit. The resulting equivalent circuit is shown in figure 7.11.

The motor characteristics are impacted by the magnitude of the source voltage and also by the source impedance. As the motor draws significantly higher currents during start-up, the terminal voltage will be reduced during this period. This reduction could be 20%–30% in some cases, and could last for times from a fraction of a second to several seconds. During starting, the motor must also develop enough torque to overcome the load torque and still provide sufficient accelerating torque.

Figure 7.12 shows the torque speed characteristic of the induction motor of table 7.1, with a source impedance of $0.012 + j0.11$ ohms and a square law load torque. At low speeds, the difference between the developed torque and the load torque goes to accelerate the motor. The accelerating torque is

$$T_{acc} = T_e - T_m \tag{7.45}$$

The motor steady state operating point will lie at the intersection of the T_e and T_m curves, when the acceleration torque is zero. In practice, the loss torque will add to the load torque to move the operating point slightly.

The accelerating torque is at least 200 N m greater than load torque from standstill up to the operating speed. This should be sufficient for starting purposes. If an accurate estimate of starting time is needed, it can be obtained by integrating the mechanical equation (7.35).

The motor draws large current during the line motor starting, which causes a voltage dip at the motor terminals. Plots of current and voltage versus rotor speed are shown in figures 7.13 and 7.14. In many applications, the voltage sag during motor starting is a critical issue, and in some cases may require starting assistance.

7.7 Squirrel cage motor performance

The two most important aspects of induction motor operation are motor starting and motor operation. A third consideration is motor protection and performance during faults.

Motor operation: A squirrel cage motor that is directly connected to the power grid will operate at a small slip just below synchronous speed. The slip will increase with motor load, and typically will be in the range of 2%–3% at full motor load. When used in 60 Hz systems, this means that induction motors will operate over a small speed range, operating just below 3600 rpm, 1800 rpm, or 1200 rpm for 2 pole,

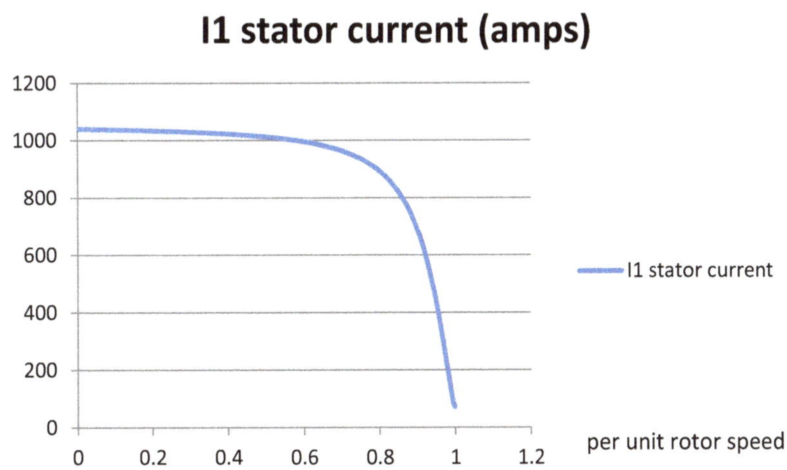

Figure 7.13. Stator phase current versus shaft speed for table 7.1 motor with source impedance.

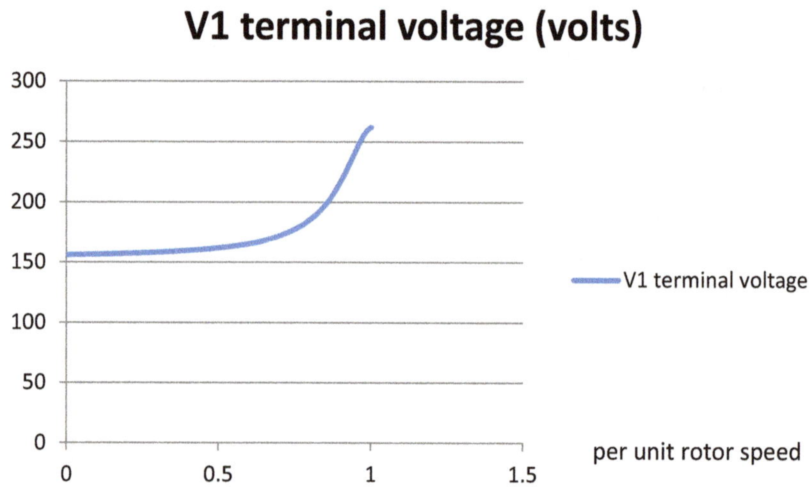

Figure 7.14. Stator line-neutral terminal voltage versus shaft speed for table 7.1 motor with source impedance.

4 pole, and 6 pole motors, respectively. These are the most common pole selections. In some applications, gearing is used to change the shaft speed of the load.

Line connected induction motors are widely used for a variety of applications in industry, commercial and residential settings. The selection of motors for most applications is based on a combination of their running capability and their ability to start successfully.

Motor running: The most straightforward applications are those with steady loads which have a steady load torque during normal operation. In these cases, the motor at least large enough to be able to meet the load torque requirement over the full range of input voltages expected. It must also be able to accelerate the motor successfully.

Variable loads include conveyor belts, fans with variable pitch blades, variable head pumps and others. In most cases, these motors will be sized in the same way as constant load motors, based on the largest expected load. There are some cases where short term overloads are allowed in the design, in cases where operation within the motor thermal limits can be maintained.

Motors driving impact loads are specified differently. A punch press is an example of an impulse load. It operates at light load for most of its operating cycle. During the punch cycle, the motor will decelerate. During this period, the press will draw energy from the power grid as well as from the kinetic energy stored in the rotating shaft. Often, these loads use specialized motors that optimize this use of energy during the short punch cycle.

Motor starting: Starting can be the most significant challenge in line connected induction motors. As discussed, the motor draws high currents and produces low torque per amp at start-up.

The power system can be represented as a constant voltage behind an impedance—essentially this is the Thevenin equivalent of the power grid. This per phase equivalent circuit is shown in figure 7.11. Across the line motor starting will cause a voltage sag at and near the motor terminals. This voltage sag can affect adjacent loads in addition to reducing the level of motor starting torque.

Motor starting tends to get more difficult as motor size increases. High starting torque requirements of the load also can create difficulties. Starting motors in an unloaded condition is helpful—unloaded conveyer belts, compressors with no head pressure, pumps with a bypass valve closed are methods that are used to aid motor starting.

For a successful start, the motor must accelerate to operating speed in a short time (typically a fraction of a second to several seconds), and must not violate short term voltage fluctuation limits—this is regularly called flicker or voltage sag. The term 'flicker' refers to the short term dimming experienced by incandescent light bulbs. Voltage flicker has been studied for many years, and flicker guidelines have been developed and incorporated into IEEE and IEC standards.

The other aspect of flicker is the location where the voltage sag is experienced. The deepest sags are experienced right at the motor terminals. In cases where neighboring customers do not experience this sag, the motor owner can choose to tolerate deeper sags rather than mitigating the motor starting problem. The power company supplying electric power to many customers will become concerned only when voltage dips due to motor starting impacting neighboring customers. This point where voltage quality impacts neighboring loads is often referred to as the point of common coupling (PCC).

Figure 7.15 shows the one line diagram and the per phase equivalent circuit this situation. It is generally sufficient to consider the voltage dips at slip s=0. In figure 7.15, the motor terminals are within a plant, and the effective impedance from the plant to the point of common coupling is $\overline{Z}_{\text{loc}}$. In most cases, the power company will enforce voltage flicker limits at the point of common coupling V_{pcc}. This is the point where other customers of the power company are affected by the voltage drop.

The impedance of the power system also has the effect of reducing the motor starting current and the starting torque. It is essential to consider the torque available at the lowest expected source voltage V_{src} and the highest expected source impedance, $Z_{\text{src}} + Z_{\text{loc}}$. The developed torque at startup must be sufficient to break the motor away from standstill. This is called the breakaway torque. It is similar to the static friction for a sliding object. For general loads, breakaway torque is typically 125%–160% of the level required to keep it turning once it has started. Certain loads can have significantly higher breakaway torques—conveyor belts with significant sliding friction are an example. The second requirement is that the motor must have sufficient accelerating torque to bring the motor up to speed in an acceptable time, before motor and system components overheat. Starting time issues most commonly arise in cases of large inertia loads.

Motor starting options: The above discussion involves motors started by direct connection to the power grid. In cases where direct connection creates voltage problems, there are several options available for starting design:

- Starting capacitors.
- Reduced voltage starting.
- Deep bar and double cage motor designs.
- Variable speed motor drive.

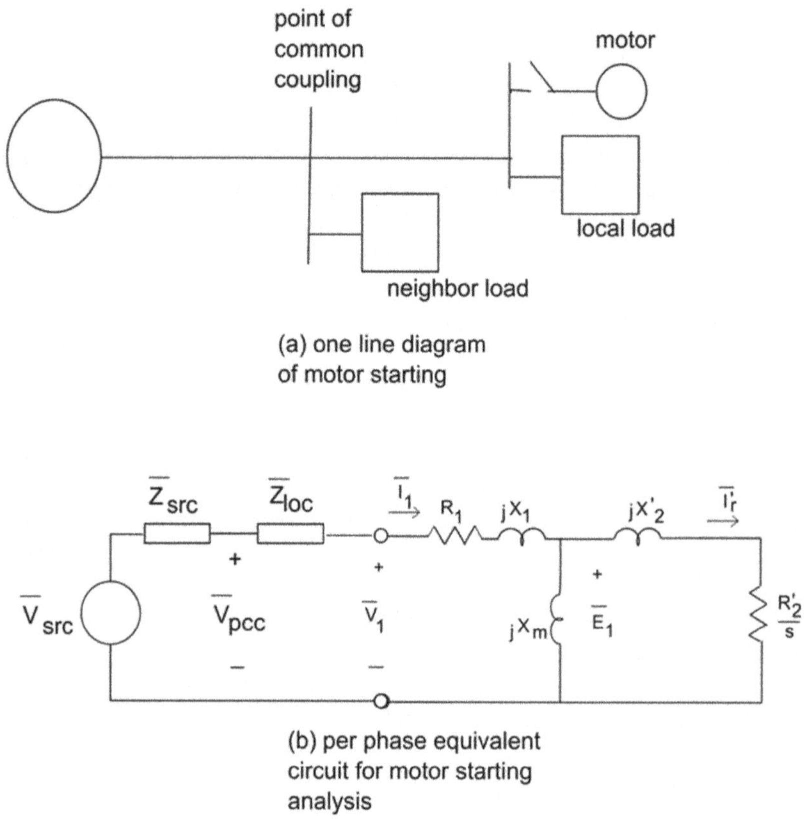

Figure 7.15. One line diagram and per phase equivalent circuit for motor starting study.

In the first case, capacitors are connected directly at the motor terminals in parallel with the motor. The capacitors counteract the inductance of the motor, and raise the terminal voltage at low slips. The starting capacitors are switched out at a shaft speed that has been determined to avoid excessive voltage dip. Large motors can also require running capacitors, to provide power factor correction and a voltage boost during normal motor operation. This method will not reduce the torque available for starting.

The second approach is reduced voltage starting. A common option currently is the power electronic 'soft start' technology. Soft start technology reduces the motor terminal voltage during starting, thereby reducing both the starting current and torque. Torque requirements for motor starting must be carefully evaluated in the application of this technology. Also, certain loads can benefit from soft start technology when it provides a smoother start then across the line starting.

Deep bar and double cage motors use the shape of the rotor bars to influence starting torque and decrease starting current. The bars on deep bar motors are rectangular or nearly rectangular with relatively high depth to width ratio. Example shapes are shown in figure 7.16. At low slips, the rotor currents are low frequency and

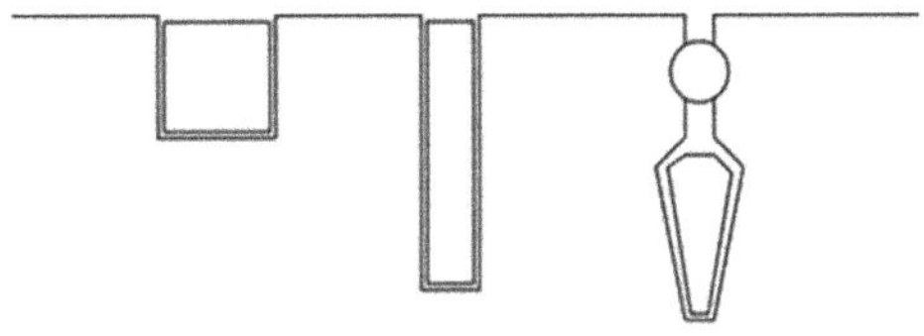

Figure 7.16. Representative examples of single cage, deep bar, and double cage rotor bars.

the current spreads uniformly through the bar. At startup, where the slip is 1.0 and the frequency of the rotor currents is the same as the source frequency, the internal inductance of the bar forces current to the surface of the rotor, which in turn causes a higher effective resistance. Double cage motors enhance this effect by using two separate cages, with the outer cage often being made with a higher resistivity metal.

In cases where these methods are not sufficient, the remaining option is to install a variable speed drive with full control of torque and current throughout motor start-up. This option also provides a wide operating speed range, which has separate benefits.

7.8 Direct connection motor starting

Many induction motors, particularly small and medium size motors with fan or pump loads, start successfully with direct across the line starting. A typical control circuit for small motors is shown in figure 7.17. The motor is controlled with start and stop pushbuttons, which energize the motor contactor (MC). A motor overload relay has thermal elements that open a contact under overload conditions, which in turns drops out the motor contactor and stops the motor. The motor is protected from short circuits by fuses that open the faulted line. Below the power circuit shown in the figure is a control logic diagram. The legend for this diagram is shown in figure 7.18. The contact of the normally open pushbutton closes when the pushbutton is pressed. Otherwise it is open. The normally closed pushbutton is the opposite—the 'normal' condition is with the pushbutton at rest, when the contact is closed. It opens on depression of the button. Similarly, the normal condition is with the coil de-energized. The normally open contact is open in this state, and closes when the coil is energized. The normally closed contact operates in just the reverse of this.

In the control diagram, when the motor is at rest, both pushbuttons and the motor contactor MC are all in their normal position. Pressing the start pushbutton energizes the contactor MC coil, closing the power contacts of the contactor as well as the auxiliary contact that is in the control circuit. This auxiliary contact is in parallel with the start pushbutton, so that when the operator releases this pushbutton, the motor contactor remains energized. This is called a seal-in contact, and it provides the 'memory' that the motor is in the ON state. When the stop pushbutton is pressed, the coil of the motor contactor is de-energized, opening the

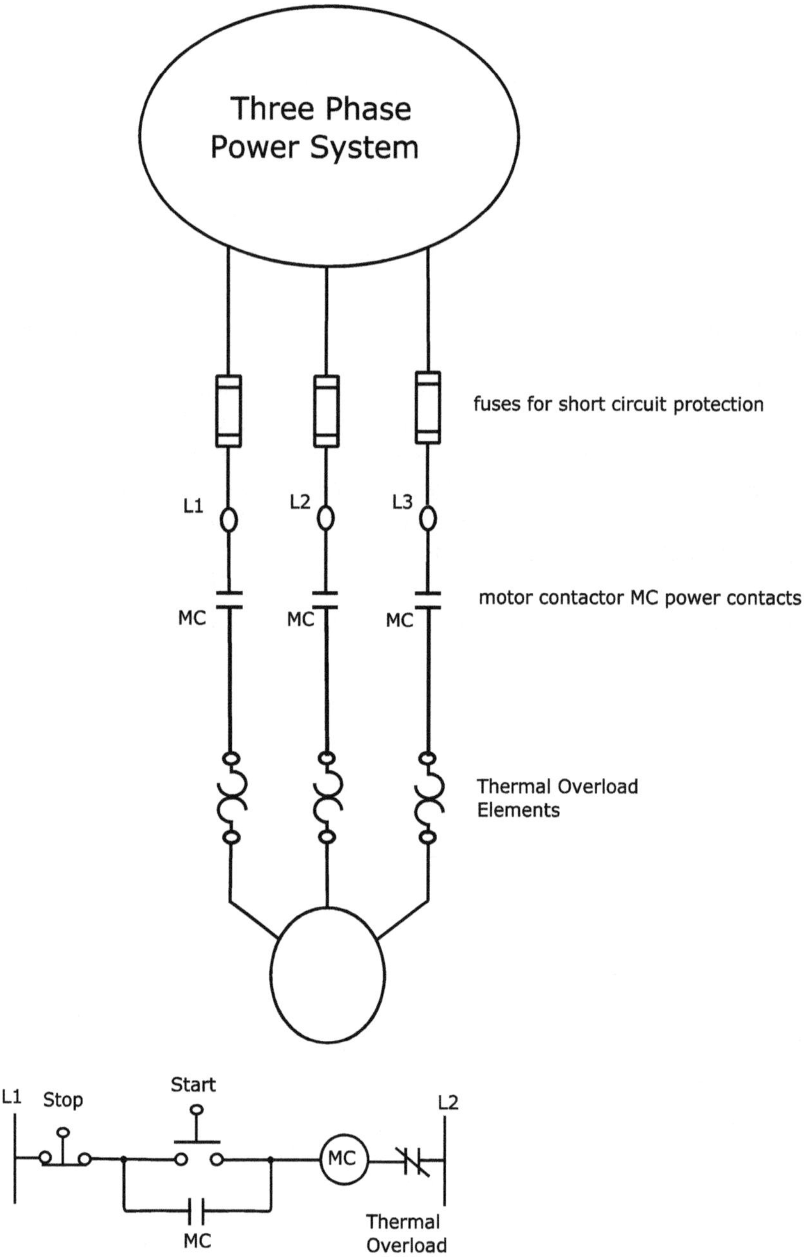

Figure 7.17. Motor starting circuit with thermal overload and fuse protection.

power and control contacts and taking the motor off line. In an overload condition, the thermal elements will cause the thermal overload contact to open, which also stops the motor. Labels 'L1' and 'L2' on the control diagram indicate that this circuit is connected to those two points on the power circuit—so the control circuit is connected across the first two phases from the source.

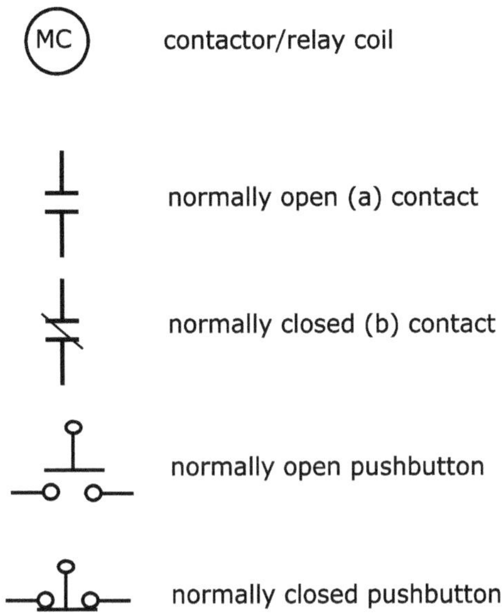

Figure 7.18. Relay logic legend.

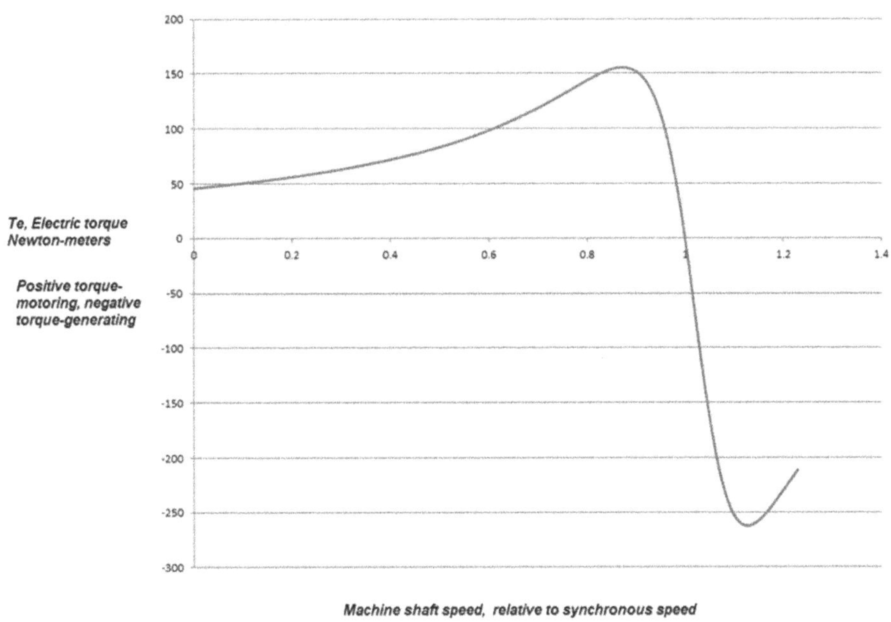

Figure 7.19. Induction machine torque–speed curve showing the generating torque when operating above synchronous speed.

Relay logic is widely applied in industry. In complex systems, the logic is often implemented in specialized computers called programmable logic controllers. These controllers are used in complex systems with many sensors, motors and actuators.

7.9 Induction generator

When squirrel cage induction machines operate above synchronous speed, slip s becomes negative, and the machine generates electric power when connected to the power grid. Directly connected induction generators are currently being used in some wind turbines, for example.

The induction generator takes its excitation from the system—just as the induction motor does. It also follows the system frequency. It therefore must be connected to a strong power grid, and cannot function as a stand-alone generator. In most cases, a VAR source will be needed at the generator site to avoid low power factor and voltage problems.

The per phase diagram of figure 7.11 can be used to analyse the induction generator. This figure uses the motor convention with current entering the machine and torque driving a mechanical load. When generating, this current will swing past −90°, the power flow at the terminal will become negative (signifying power flow out of the motor). The developed torque will also be negative, and the 'load' torque will also be negative when using the motor convention. The resulting torque speed curve is shown in figure 7.19, where machine operation above synchronous speed produces a negative torque in the motoring convention.

Example 7.6. The induction machine of table 7.1 is connected to the power grid as shown in figure 7.11. The source impedance is $Z_{\text{th}} = 0.012 + j0.11\Omega$, and the system voltage is $\overline{V}_{\text{th}} = 265\text{V}$. The system frequency is 60 Hz. The machine is operating as a generator, with a shaft speed of 1840 rpm.

Find the stator current for this operating point, and the generator developed torque.

The slip at 1840 rpm is

$$s = \frac{1800 - 1840}{1800} = -0.0222$$

The rotor impedance is then

$$\overline{Z}_r = \frac{0.026}{-0.0222} + j0.072 = -1.485 + j0.072\Omega$$

The total impedance of the motor plus source is

$$\overline{Z}_{\text{total}} = 0.012 + j0.11 + 0.021 + j0.072 + \frac{j3.6(-1.485 + j0.072)}{-1.485 + j3.672}$$

$$= -1.194 + j0.749\Omega$$

The stator current is then

$$\overline{I}_1 = 188.1 \underline{/-148.4^\circ} \text{ A}$$

Using the current divider, the rotor current is

$$I_r' = 170.9 \underline{/-170.5^\circ} \text{A}$$

From equation (7.38), the torque is

$$T_e = \frac{3(\text{poles})}{2\omega_e} \frac{I_r'^2 R_2'}{s} = \frac{3 \cdot 4}{2 \cdot 377} \frac{170.9^2 \cdot 0.026}{-0.0222} = -544.4 \text{ N m}$$

Find the terminal real and reactive power.

$$\overline{V}_1 = \overline{V}_{th} - (R_{th} + jX_{th})\overline{I}_1 = 256.6 \underline{/4.2^\circ \text{v}}$$

The real power flow is then

$$P_1 = 3 \cdot 256.6\text{V} \cdot 188.1 \cos(4.2^\circ + 148.4^\circ) = -128.6 \text{ kW}$$

The real power is –128.6kW entering the machine, so it is 128.6kW generated by the machine and entering the power grid.

The VARs entering the machine are

$$Q_1 = 3 \cdot 256.6\text{V} \cdot 188.1 \sin(4.2^\circ + 148.4^\circ) = 66.6 \text{ kVAR}$$

Being positive, these are VARs entering the machine. As stated earlier, VARs must enter this machine in both motoring and generating operation, as the VARs are required to provide the machine excitation.

Questions

1. Search the internet for photos or illustrations of a squirrel cage, and sketch a view of the rotor bar assembly and end rings.
2. The National Electric Manufacturer' Association (NEMA) has developed five classes of induction motors that are suitable for different types of motor loads. These different design classes influence the rotor bar design, as illustrated in figure 7.12. From a web search, determine the typical applications for NEMA Class B and Class D motors.
3. Induction motors are used in a wide variety of applications. Some of these applications involve their placement in hazardous environments. A set of motor classes has also been developed for application in a range of hazardous locations. Research explosion proof motors, and describe the characteristics of these motors and the environments that these motors are designed for.
4. Section 7.7 discusses voltage sags and flicker. Do a web search on 'flicker curve.' Discuss the tolerable limits of flicker that you find, and cite the source of the flicker data.

Problems

1. A 75 horsepower squirrel cage induction motor has the following characteristics:

$R_1 = 0.076\Omega$	$X_1 = 0.19\Omega$	$X_m = 13.3\Omega$	$f_1 = 60\text{Hz}$	phases = 3
$R'_2 = 0.08\Omega$	$X'_2 = 0.19\Omega$	poles = 4	$R_1 = 0.076\Omega$	

The motor is operating at a speed of 1755 rpm with a terminal voltage of 266 V line to neutral (460 V L-L).
   a. Calculate the motor slip and the rotor frequency f_2.
   b. Calculate the stator current $\overline{I}_1$ and the rotor current $\overline{I'}_r$. Assume that the motor terminal voltage is at 0° for this calculation.
   c. Calculate the real and reactive power flowing into the motor terminals.
   d. Calculate the developed torque of the motor T_e.
2. This same motor is fed from a source with an open circuit voltage of 277 V and a source impedance of $\overline{Z}_{th} = 0.04 + j0.11\Omega$. Determine the motor terminal voltage, stator current and torque at motor starting ($s = 1$).
3. Plots of developed torque and load torque versus speed (such as are shown in figure 7.12) are useful in determining motor operating point, acceleration torque and other operating information. These can be developed by repetitively solving the steady state circuit of figure 7.11 over the range of speed from standstill to synchronous speed. The Excel file induction_motor.xls (available from http://iopscience.iop.org/book/978-0-7503-1662-0) does these calculations and plots. Run this file for the motor of question 2, and determine the following:
   a. The peak torque that will be developed by the motor, and the motor slip at peak torque.
   b. The operating speed of the motor when the load is a square load torque with 15 N m torque at standstill and 300 N m of torque at synchronous speed.
   c. Repeat (a) and (b) when the source voltage drops by 5%. What is the percentage impact of this voltage drop on peak torque, and on motor slip?
4. A doubly fed induction generator has the following parameters:

R_1	L_1	L_m	L_2	R_2	Turns ratio a	Poles	Frequency
0.06 ohms	0.849 mH	250 mH	0.256 mH	0.00925 ohms	2	4	60 Hz

Use the per phase equivalent circuit shown in figure 7.3 to solve this problem, with the current directions reversed for generator convention.
   a. The generator is operating at 1700 rpm. What is the slip of the generator?

b. The generator is operating with a stator line to neutral voltage of $\overline{V}_1 = 2400\text{V}/\underline{0^\circ}$. The stator winding is delivering 1 MW at unity power factor to the grid. Determine the value of the current $\overline{I}_1$.
c. Solve the per phase equivalent circuit to find the rotor voltage and current.
d. Find the real and reactive power flowing out of the rotor.
e. Repeat for a generator shaft speed of 1900 rpm

IOP Publishing

Electromechanical Machinery Theory and Performance

Thomas Ortmeyer

Chapter 8

Power electronic converters and speed control of AC machines

There are numerous motor and generator applications that require variable speeds. Examples include electric vehicles and wind turbines. Historically, variable speed drives used DC machines, with the power supplied by a DC generator driven by an induction motor. As technology advanced, the motor-generator set was replaced by power electronic controlled rectifiers that supplied a variable DC voltage to the motor. More recent advances have led to the development of electronic rectifier/inverters that feed AC motors. These modern AC variable speed drives have lower cost and improved performance for many of the variable speed applications.

8.1 Pulse width modulated converters: the full bridge converter

The majority of AC drives use pulse width modulated (PWM) converters with insulated gate bipolar transistor (IGBT) switches. While other converter types and switch technologies exist, they are not covered in this book.

The IGBT is a hybrid transistor that is controlled by a gate voltage that uses field effect transistor (FET) technology. The IGBT is a two stage device, with the FET drive then enabling a bias junction transistor (BJT) output stage that alternately conducts or blocks the flow of current. The BJT output stage provides low conduction loss as well as good voltage blocking characteristics. The FET input stage provides a significant improvement in device drive requirements and also in device speed. IGBTs are currently available with current ratings in the kiloamp range and voltage ratings in the kilovolt range. There are several different device symbols in use for the IGBT, one of these is shown in figure 8.1.

The characteristic curves for a typical IGBT are shown in figure 8.2. This curve shows plots of collector (C) current versus collector to emitter voltage for a range of gate to emitter voltages. This device is intended to operate in either the 'on' state or the 'off' state. In the on state, the gate to emitter voltage V_{GE} is applied sufficiently

doi:10.1088/978-0-7503-1662-0ch8

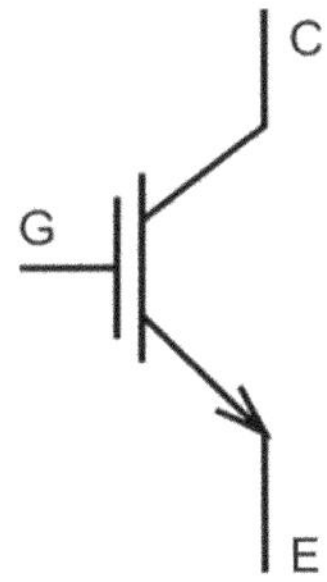

Figure 8.1. IGBT symbol (one of several).

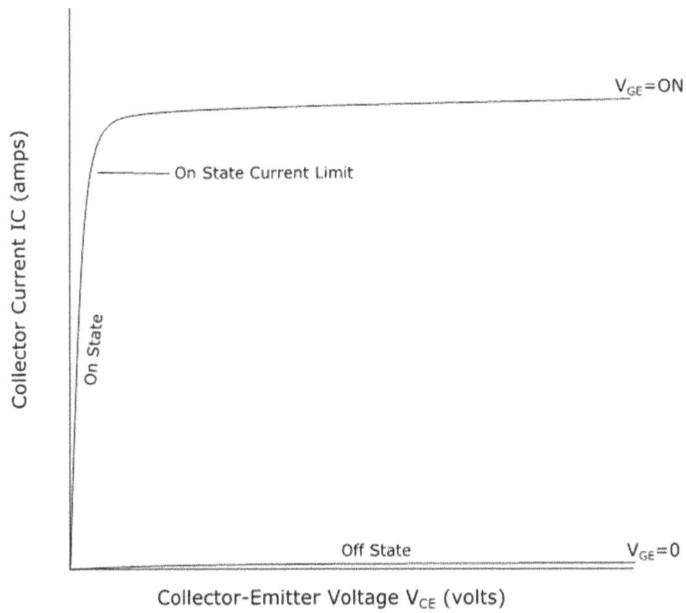

Figure 8.2. Typical on state and off state characteristic curves for power IGBTs. The IGBT experiences high conduction loss if operated above the on state current limit.

large so that the IGBT conducts current in the saturation region, with minimum voltage drop. In the off state, $V_{GE} = 0$ V, and the IGBT blocks conduction. In this state, it operates very close to the horizontal axis with currents typically in the sub-milliamp range. When switching between states—both on to off and off to on, the device moves through the active region. Device power losses are therefore higher during switching, and these switching losses are an important factor in the overall converter performance and rating.

Notice that the characteristic curve is strictly in the first quadrant. IGBTs do not block current in the reverse direction—essentially they have an inherent diode connected anti-parallel from emitter to collector. In applications with reverse current flows, the IGBT is generally drawn with this diode shown. Physically, the diode can

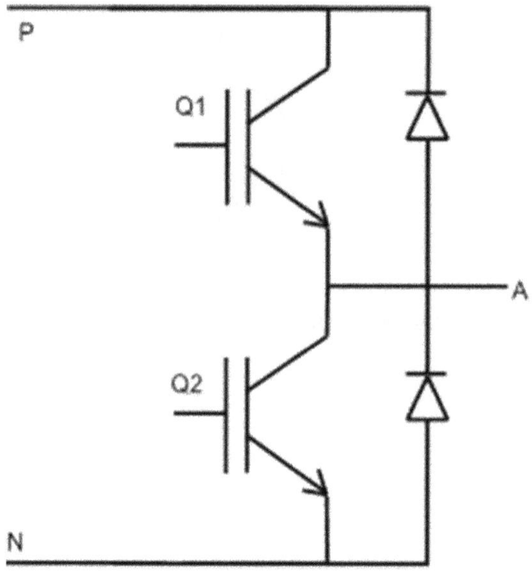

Figure 8.3. Power pole circuit of two IGBTs with anti-parallel diodes.

be an integral part of the IGBT, or it can be a separate device, perhaps on the same substrate as the IGBT.

In many power applications, two IGBT/diode pairs are connected in series between a positive and negative DC bus. This connection is often called a power pole. The power pole circuit is shown in figure 8.3. In the power pole arrangement, the P and N terminals are connected to a DC circuit, and the A terminal is connected to an AC circuit. Multiple power poles can be configured in a given converter to provide a single phase or three phase AC source. These circuits are often referred to as inverters, which implies power flowing from a DC source to an AC load. These circuits can operate with power flow from AC to DC as well, so the name 'converter' is a more accurate description than 'inverter.' For this reason, the term converter is used in this text.

8.1.1 DC to DC PWM converter

A two power pole circuit is shown in figure 8.4. This circuit will perform both DC to DC power conversion and DC to single phase AC power conversion. With the reference directions shown, this converter is assumed to be converting power from the DC source V_{in} to the output load circuit. The input voltage V_{in} must be a positive DC voltage, while voltage V_{AB} can be either positive or negative, with a magnitude less than the voltage V_{PN}. The currents I_{in} and I_A can be either positive or negative.

This circuit is also referred to as a full bridge circuit. Note that, if the two switches in either power pole are on at the same time, the source voltage will be short circuited. Therefore, an operating requirement for this bridge is that Q_2 must be off whenever Q_1 is on, and Q_1 must be off whenever Q_2 is on. Similar rules apply for switches Q_3 and Q_4.

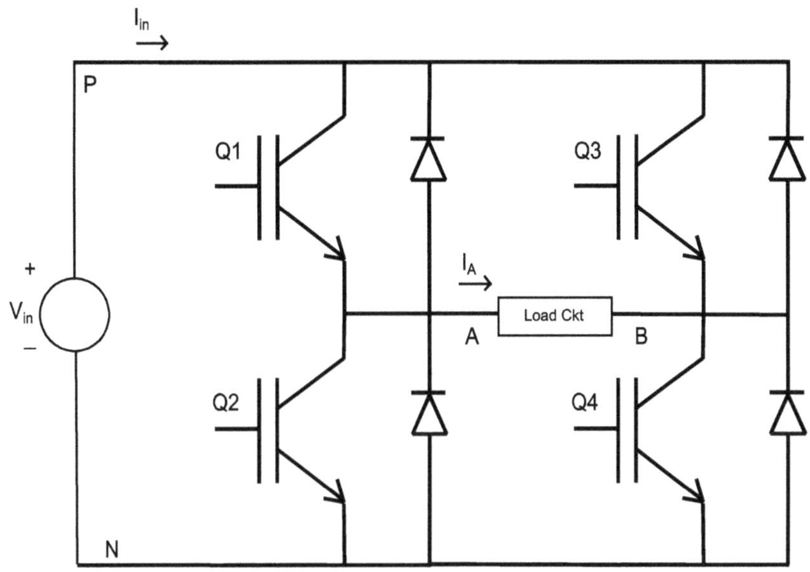

Figure 8.4. DC to DC or DC to single phase AC PWM converter.

Table 8.1. Two state operation of the full bridge.

	On state	Off state	V_{AB}
State 1	Q_1, Q_4	Q_2, Q_3	$+V_{in}$
State 2	Q_2, Q_3	Q_1, Q_4	$-V_{in}$

In its simplest operating mode, the switches are operated in two states, as shown in table 8.1.

In this operating mode, one and only one switch is on per power pole. (Note: in practice, a short interval is required between turning off one of these switches and turning on the other.) When on, the switches act as short circuits, and when off, the switches act as open circuits. Therefore, in State 1, the input voltage V_{in} is directly connected across the load $V_{AB} = V_{in}$. In State 2, the input voltage is again connected to the load, but in the reverse direction, $V_{AB} = -V_{in}$.

Assuming that switching between states is fast, this converter will be either in State 1 or State 2 at any given instant. If this bridge is switched repetitively between State 1 and State 2 with constant on time values for both states, the output voltage will have a constant DC component plus harmonics at and above the switching frequency. If the bridge is in State 1 for t_1 seconds, and State 2 for t_2 seconds, the total switching period will be

$$T = t_1 + t_2 \tag{8.1}$$

In this type of bridge, State 1 is often considered the 'on' state, and State 2 the 'off' state. The duty cycle is defined as the ratio of the on state time to the total period,

$$D = \frac{t_1}{T} \tag{8.2}$$

The resulting output waveform is shown in figure 8.5. From this figure, the DC (average) value of the PWM waveform is

$$V_{AB} = \frac{1}{T}[DTV_{in} - (1-D)TV_{in}] = (2D-1)V_{in} \quad \text{for } 0<D<1 \tag{8.3}$$

For convenience, a modulation factor m is often defined, where

$$m = (2D-1) \tag{8.4}$$

Note that, as the range of D is 0 to 1, the range of m is −1 to 1. In terms of modulation factor, the average output voltage is then

$$V_{AB} = mV_{in} \tag{8.5}$$

This equation shows that V_{AB} can range continuously from $-V_{in}$ to $+V_{in}$ as the duty cycle is varied from 0 to 1. The duty cycle is set by the bridge controller.

The switching frequency for this bridge is the inverse of the period,

$$f_s = \frac{1}{T} \tag{8.6}$$

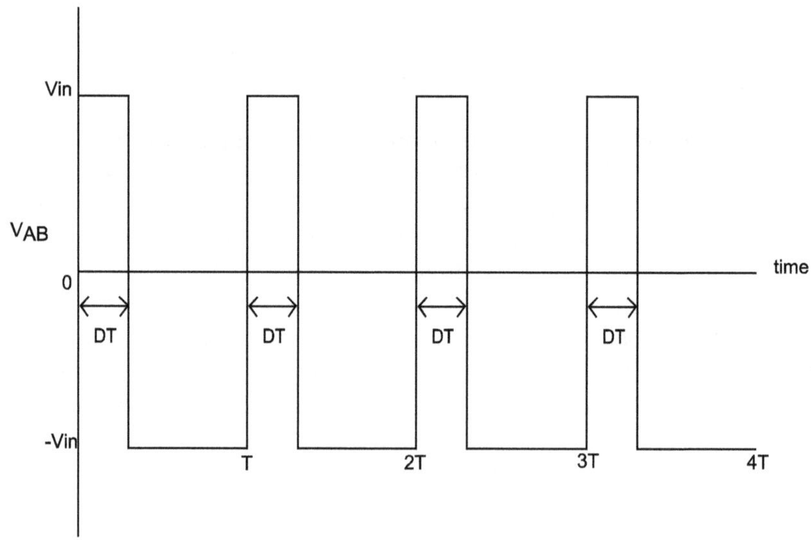

Figure 8.5. PWM waveform for duty cycle $D = 0.3$.

For IGBT switches, typical switching frequencies are in the kilohertz to tens of kilohertz range. Small switching power supplies that use FETs as switching can switch at and above 100 kHz.

Figure 8.5 shows that there will be significant ripple in the output voltage in addition to its DC component. The load circuit for this converter will typically include inductance that will reduce the level of current ripple. For example, the model for the armature winding of a DC motor is shown in figure 8.6. The naturally occurring inductance of this circuit is often sufficient to reduce current ripple to an acceptable level, so that additional filtering is not generally needed in DC motor drive applications.

Example 8.1. Consider the full bridge converter shown in figure 8.4. Figure 8.5 shows the converter operation in DC–DC bipolar mode. An alternate control mode for full bridge DC–DC converters is unipolar mode. This converter operates in unipolar mode if the switches are controlled as shown in table 8.2. As shown in the table, there are periods when switches 1 and 3 are on, and periods when switches 2 and 4 are on. In both cases, the output voltage $V_{AB} = 0$.

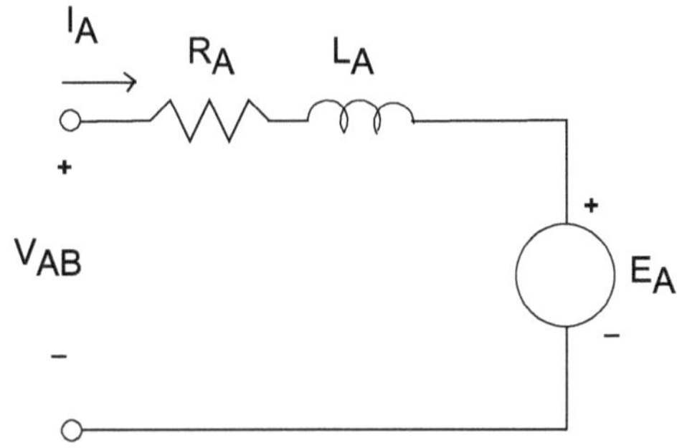

Figure 8.6. Equivalent circuit of DC motor armature.

Table 8.2. Two switching cycles of unipolar DC–DC converter.

Cycle #	Time range (μs)	On switches	Off switches
1	0–20	1,4	2,3
1	20–60	1,3	2,4
1	60–80	1,4	2,3
1	80–120	2,4	1,3
2	120–140	1,4	2,3
2	140–180	1,3	2,4
2	180–200	1,4	2,3
2	200–240	2,4	1,3

(a) Sketch the output voltage V_{AB} over the 240 μs shown in the table.

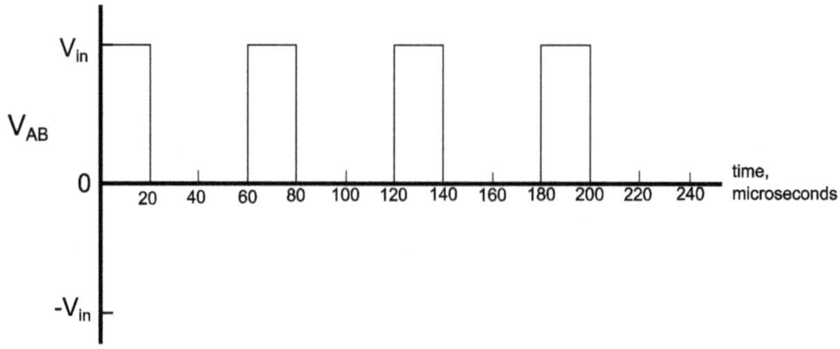

(b) Table 8.2 shows two full switching cycles. Analyse the output waveform.

- The output waveform has a period of 60 μs.
- Each switch goes through one cycle of on–off operation with a period of 120 μs. However, the output voltage has a period of 60 μs.
- The output waveform has a duty cycle $D = 0.33$.
- The output voltage never goes negative, and has lower ripple content.
- The output voltage is $V_{AB} = DV_{in}$.

Unipolar operation has several advantages over the bipolar operation that is shown in figure 8.5.

8.1.2 DC to single phase AC PWM converter

The bridge circuit shown in figure 8.4 is also used as a DC to single phase AC power converter. In order to achieve AC output from this circuit, the duty cycle D is continuously varied. When this variation is slow relative to the PWM switching frequency, the resulting performance is good. A ratio of switching frequency to output fundamental frequency greater than 20 is often considered suitable for good operation.

Sinusoidal pulse width modulation is a common method for controlling the amplitude and frequency of the bridge output voltage $v_{AB}(t)$. When the modulating factor m is

$$m = m_A \sin(2\pi f_1 t) \qquad -1 < m < 1 \tag{8.7}$$

The fundamental value of the output voltage of the bridge is then

$$v_{AB}(t) = mV_{in} = m_A V_{in} \sin(2\pi f_1 t) \tag{8.8}$$

In this equation, m_A is the amplitude modulation ratio, a value between 0 and 1. It is a constant for a given output amplitude sinusoid.

A *frequency modulation* ratio is defined as

$$m_{\mathrm{f}} = \frac{f_{\mathrm{s}}}{f_1} \tag{8.9}$$

A common way to implement sinusoidal pulse width modulation (SPWM) is shown in figure 8.7. In this method, a control signal is compared with a triangular carrier signal. The carrier signal frequency f_{s} defines the switching frequency of the bridge, and the frequency of the control signal defines the fundamental frequency f_1 of the bridge output. When the control signal is greater than the carrier signal, the bridge is operated in State 1. When the control signal is less than the carrier signal, the bridge is operated in State 2. The resulting output voltage $v_{\mathrm{AB}}(t)$ is shown in the figure. The amplitude modulation ratio is simply the ratio of the peak amplitude of the control signal to the peak amplitude of the triangular carrier signal.

In practice, these bridges are operated with much higher frequency modulation ratios. The resulting output waveform consists of a fundamental component and a set of harmonics that are at and above the switching frequency. Load circuit inductance is used to limit the harmonic current flow to acceptable levels.

Some full single phase bridges are operated with four states—the two states of table 8.1, and two additional states, State 3 with Q1 and Q3 on and Q2 and Q4 off, and State 4 with Q2 and Q4 on and Q1 and Q3 off. This operating mode provides

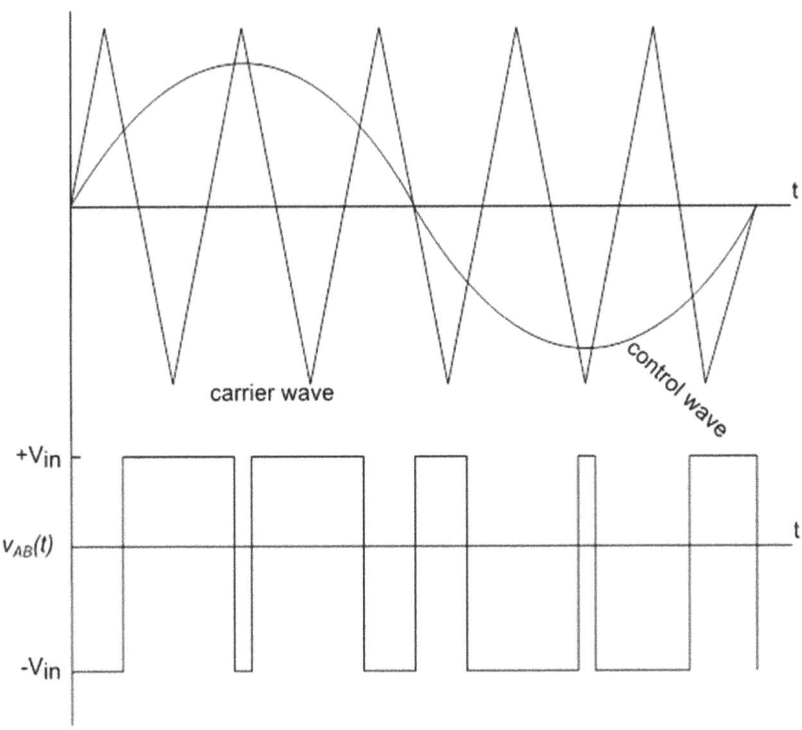

Figure 8.7. Example control and output voltage waveforms for SPWM with $m_A = 0.8$ and $m_f = 5$.

the same fundamental output voltage, but has reduced harmonic levels as compared to the bridge control strategy shown in figure 8.7.

Example 8.2. Consider the full bridge converter shown in figure 8.4. The converter is operating as a DC–AC inverter. The fundamental value of the inverter output voltage is given by equation (8.8).

(a) If $m_a = 0.80$, $V_{in} = 240$ V, and $f_1 = 60$, find the instantaneous value of the fundamental output voltage.

From equation (8.8), $v_{AB}(t) = 0.8 \cdot 240\sin(2\pi 60t) = 192\sin(377t)$V

(b) Find the rms value of the fundamental voltage.

$$V_{AB} = \frac{192\ \text{V}}{\sqrt{2}} = 135.8\ \text{V}$$

(c) The inverter is loaded by the circuit shown in figure 8.6, with $R_A = 20\Omega$, $L_A = 25$mH, and $E_A = 0$. Find the amplitude and phase angle of the fundamental load current.

Assume that the fundamental voltage angle is 0°. The load impedance is

$$\overline{Z}_A = R_A + j\omega L_A = 20 + j377(0.025) = 20 + j9.43\ \Omega$$

The fundamental (60 Hz) current is then

$$\overline{I}_A = \frac{135.8\ \text{V}\ \underline{/0°}\ \text{V}}{20 + j9.43\ \Omega} = 6.14\underline{/-25.2°}\text{A}$$

(d) The lowest harmonic voltage is measured to be the 49th harmonic, and has an rms amplitude of 93 V. Determine the amplitude of the 49th harmonic current.

$$Z_{AB49} = |20 + j(49)(2\pi 60)(0.025)| = |20 + j462| = 462.2\ \Omega$$

The *amplitude* of the current is then

$$I_{AB49} = \frac{93\ \text{V}}{462.2\ \Omega} = 0.201\text{A}$$

Even though the fundamental and the 49th harmonic voltages have similar amplitudes, the 49th harmonic current is significantly smaller than the fundamental current, due to the higher harmonic impedance of the load.

8.2 Three phase PWM converter

The simplest and most common three phase PWM converter bridge is shown in figure 8.8. This bridge simply adds one additional power pole to the full bridge

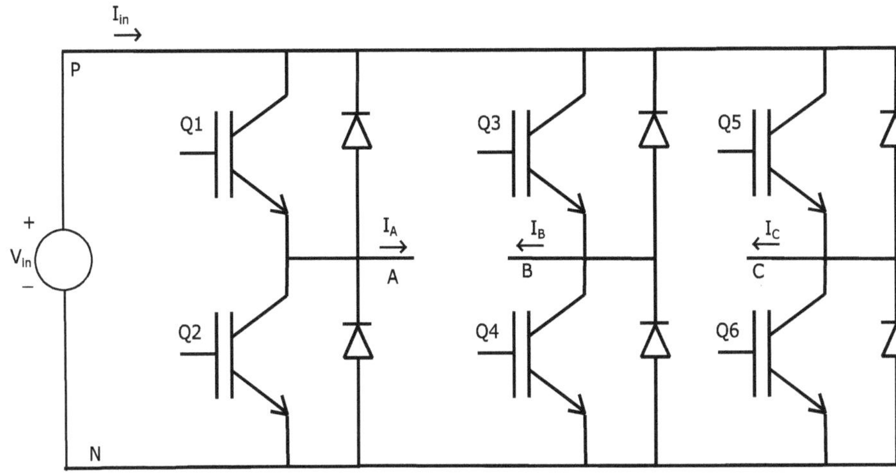

Figure 8.8. Three phase PWM bridge.

PWM converter of section 8.1. A three phase load is connected to points A, B, and C of the bridge.

Three phase bridges of this type are generally controlled either by a three phase form of sinusoidal pulse width modulation, or by a technique called space vector modulation. Space vector modulation provides somewhat better performance. In modern drives, both of these control schemes are implemented digitally, using microcontroller, DSP, or FPGA technologies.

Sinusoidal pulse width modulation

The sinusoidal PWM technique uses a triangular carrier wave, the same as the carrier wave used in single phase converters. A sinusoidal control wave is used for each leg of the inverter. Figure 8.9 shows these waves for an example case, with a frequency modulation ratio $m_f = 10$, and an amplitude modulation ratio $m_a = 1.0$.

The resulting command logic signal trace for the A phase leg of the converter is shown in figure 8.10. The instantaneous voltage between the phase A output of the bridge and the negative bus on the bridge input will be this logic signal times the bridge input voltage $V_{in.}$ In this figure, it can be seen that this signal is not quarter wave symmetric. As a result, it will have some even harmonic frequencies. When the frequency modulation is sufficiently high, these non-standard harmonic frequencies become negligible, and the frequency modulation ratio does not need to have an integer value.

Space vector pulse width modulation

In space vector pulse width modulation (SVPWM), each of the three power poles is assigned a logic value of either 0 or 1. A three digit vector of these logic states then defines the state of the inverter. For example, the logic vector 010 states that the A phase inverter leg is in state 0, meaning switch 1 is off and switch 4 is on. The B phase leg is in state 1, meaning that switch 3 is on and switch 5 is off, and in the C phase leg, the state is 0, so switch 5 is off and switch 6 is off. As a result, the A and C phases are connected to the negative bus, and the B phase is connected to the positive bus.

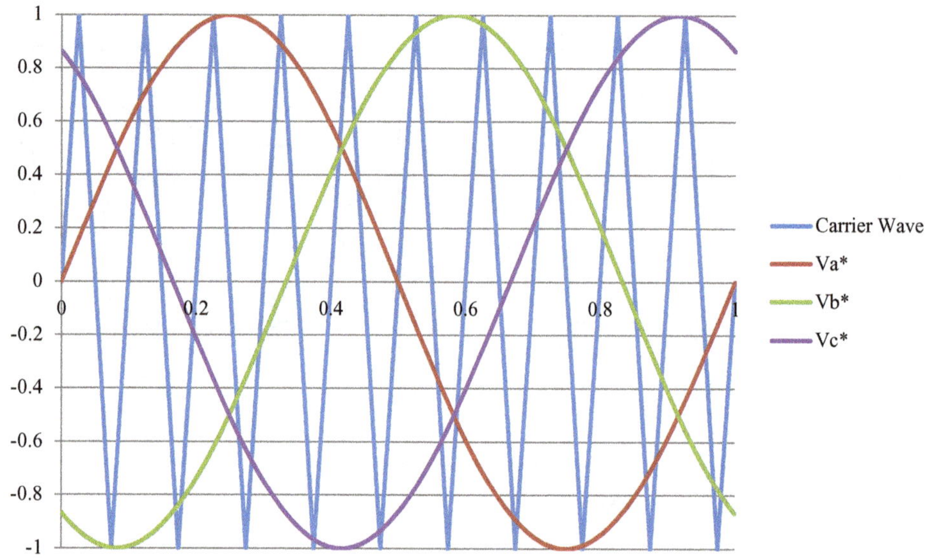

Figure 8.9. Carrier and control waveforms for three phase SPWM. Frequency modulation ratio $m_f = 10$, and an amplitude modulation ratio $m_a = 1.0$.

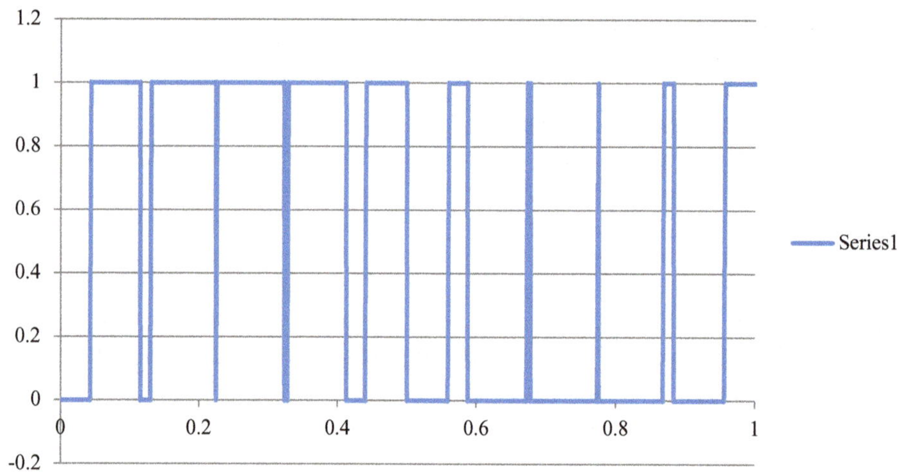

Figure 8.10. Gate command logic signal for A phase switch 1 for the SPWM case of figure 8.14.

As a result,

$$V_{AB} = -V_{in}$$
$$V_{BC} = +V_{in}$$
$$V_{CA} = 0$$

Table 8.3 shows the eight possible states that the bridge operates in, with the resulting voltages.

Table 8.3. The eight switching states for space vector modulation (SVM).

State	Switch state vector	V_{AB}	V_{BC}	V_{CA}
0	000	0	0	0
1	100	$+V_{in}$	$-V_{in}$	0
2	110	0	$+V_{in}$	
3	010	$-V_{in}$	$+V_{in}$	0
4	011	$-V_{in}$	0	$+V_{in}$
5	001	0	$-V_{in}$	$+V_{in}$
6	101	$+V_{in}$	$-V_{in}$	0
7	111	0	0	0

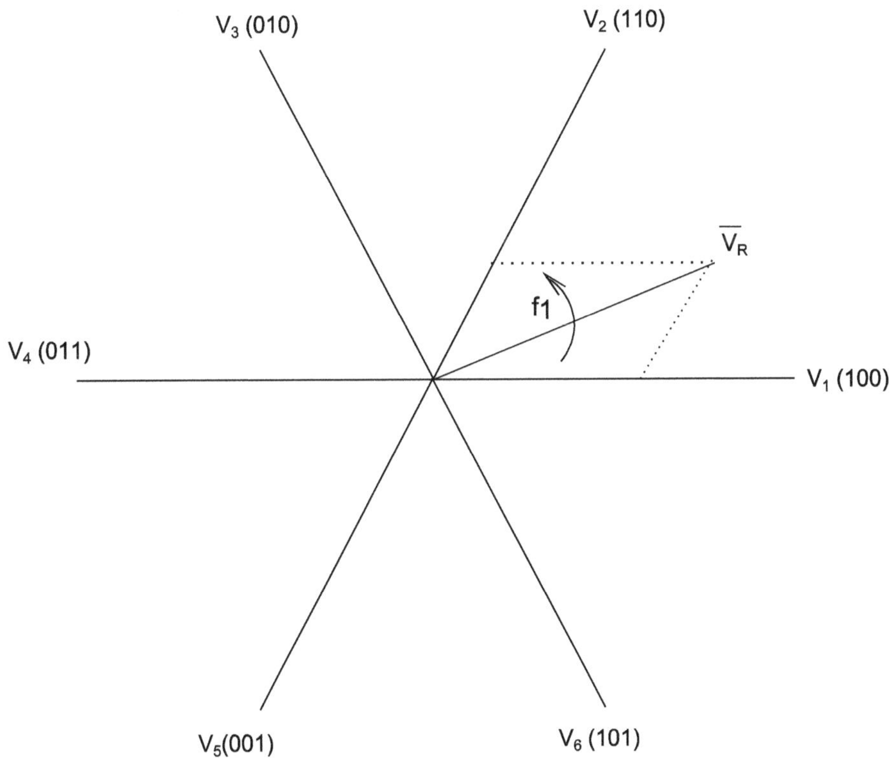

Figure 8.11. Vector diagram for space vector pulse width modulation

Figure 8.11 shows the vector diagram for implementing SVPWM. Six stationary vectors are defined, and a switching state is associated with each vector. Each vector has the same length, and the vectors are located with 60° spacing as shown. A single control or reference space vector $\overline{V}_R$ is then defined. $\overline{V}_R$ is rotating at f_1 rotations per second, and has amplitude V_R. As a result, the frequency of the inverter bridge will be

f_1 hertz, and the amplitude of the bridge output voltage will be proportional to the magnitude V_R. In figure 8.11, $\overline{V}_R$ is shown in the first sector, between V_1 and V_2. For a given position of $\overline{V}_R$, the bridge will operate in state V_1 for the proportion of time associated with the projection $\overline{V}_R$ of onto V_1, and it will operate in state V_2 for the proportion of switching period time associate with the projection of $\overline{V}_R$ onto V_2. For any remaining time in the switching period, the bridge will be in one of the null states 000 or 111.

In the next switching period, the angle of $\overline{V}_R$ will advance by an angle set by the output frequency f_1 and the period of the switching frequency. The proportion of times in each state will be adjusted according to the new position of $\overline{V}_R$, and these new times will be implemented on the bridge. At the point where the vector $\overline{V}_R$ passes from one sector into the next, the bridge will go into the new switching mode set by the stationary vectors it is then between.

It should also be noted that in motor drives, the amplitude V_R and speed f_1 are variable quantities set by the drive controller.

Equivalent circuit representation

It is often convenient to show the three phase PWM converter as a 'black box' when studying the input/output performance of the bridge. This diagram is shown in figure 8.12, with a three phase load connected to the output phases. The particular load shown in this figure is a balanced wye connected series RL load. In practice, any three phase load can be connected to the output of this bridge. Note that in this diagram the neutral point of the three phase circuit is floating, and ground has been connected at the midpoint of the DC source. There is significant voltage between this AC neutral and the DC input to the bridge. In practice, a bridge of this type will have a single ground at some point chosen by the designers. This could be, for example, at the negative bus of the DC supply, the neutral point of the AC load, or

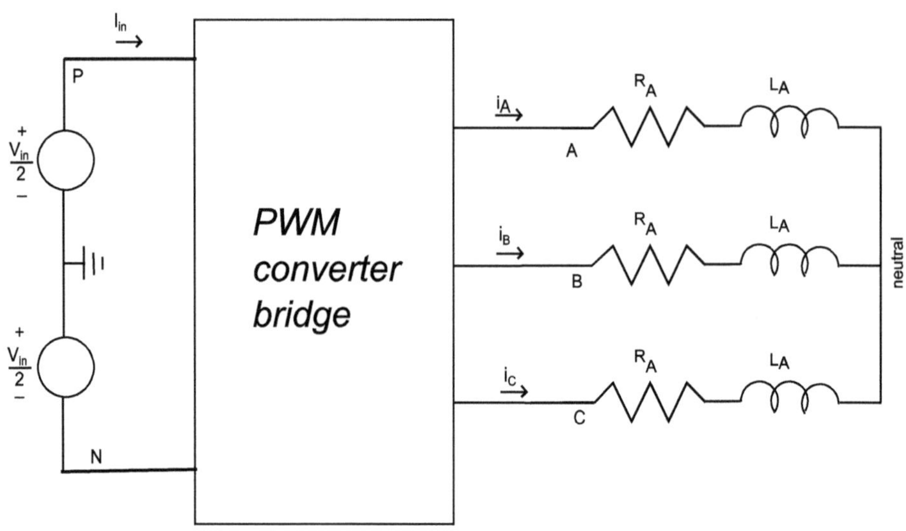

Figure 8.12. Three phase PWM bridge as a block diagram, showing the DC input and AC output quantities. The load in this diagram is a series RL load with no internal emf.

this center point of the DC source. This ground is for safety purposes, and will not carry any load current. It may, however, carry high frequency currents that result from stray capacitance to ground.

In well-designed applications, the harmonic content of the drive can be neglected for steady state fundamental frequency analysis of the input/output power flow and torque production. Harmonics can create additional losses and torque ripple, and need to be considered in the overall design of drive systems.

An example of three phase sinusoidal pulse width switching logic is shown in figures 8.9 and 8.10. This example shows an amplitude modulation ratio of $m_A = 1$, signifying that the control wave and carrier wave have the same magnitude. This is the largest value of m_A that is typically used in practice, as higher values result in undesirable low order harmonic current flow.

With $0 \leq m_A \leq 1$, the fundamental A phase to neutral voltage in the three phase bridge will be (with the subscript 'n' referring to the neutral point of the three phase system):

$$v_{An}(t) = \frac{m_A}{2} V_{in} \sin(2\pi f_1 t) \tag{8.10}$$

The B and C phase voltages will have equal magnitudes, and be shifted by $2\pi/3$ radians or 120°, so that

$$\begin{aligned} v_{Bn}(t) &= \frac{m_A}{2} V_{in} \sin(2\pi f_1 t - 2\pi/3) \\ v_{Cn}(t) &= \frac{m_A}{2} V_{in} \sin(2\pi f_1 t - 4\pi/3) \end{aligned} \tag{8.11}$$

This set of voltages could, of course, be at some reference angle other than at 0° in A phase.

The RMS line to neutral voltage magnitude will be

$$V_{An} = \frac{m_A}{2\sqrt{2}} V_{in} = 0.354 m_A V_{in} \tag{8.12}$$

The magnitude of the RMS line to line voltage will then be

$$V_{AB} = \sqrt{3}\, V_{An} = 0.612 m_A V_{in} \tag{8.13}$$

These equations are good for amplitude modulation ratios $0 \leqslant m_A \leqslant 1.0$.

Due to the phase shift involved in modulating A, B and C phase power poles, the peak of the fundamental value of the line to line voltage never reaches a value of V_{in}. There are several methods for increasing the peak value to V_{in}, of which the space vector method is currently the most common. Using any of these methods, a peak instantaneous fundamental output voltage of V_{in} can be realized, which is equivalent to an RMS line to line voltage of $0.707 V_{in}$. For a given value of V_{in}, this represents a 15% increase in maximum output voltage capability of the bridge, as compared with equation (8.13) for sinusoidal PWM. This increase in fundamental voltage output is realized solely with an upgrade of the control algorithm, and with no change in

IGBT ratings of the inverter. In the notation of this book, the *fundamental* RMS line to neutral and line to line voltages for space vector modulation are:

$$V_{An} = 0.41 m_A V_{in}$$
$$V_{AB} = 0.\ 707 m_A V_{in} \quad \textit{for}\ 0 \leqslant m_A \leqslant 1.0 \tag{8.14}$$

Note that, while the notation on these voltages refers to the A phase and A to B quantities, the magnitudes are the same for the other two line to neutral and line to line voltages. Also, it is necessary to emphasize that these quantities are the RMS or effective values of the fundamental only. The total RMS value of these voltages will be substantially higher than these values. However, it is the fundamental component that does the useful work in these systems, and it is appropriate and useful to focus on the fundamental quantities.

Three phase synchronous and induction machines present balanced loads/sources to the converter circuit. Steady state analysis of these machines can be done using the per phase equivalent circuit, similar to the analysis that was done with these machines connected directly to the electric power grid. The difference between these two applications is that the converter provides controllable frequency and voltage output that varies over a wide range.

Example 8.3. For the circuit shown in figure 8.12, $V_{in} = 550\text{V}$, $R_A = 8\Omega$, and $L_A = 2.3\text{mH}$. The inverter is operating with sinusoidal pulse width modulation. When $m_a = 0.60$ and $f_1 = 90$ Hz, find the A phase line to neutral voltage, A phase current flow, and the three phase watts and VARs flowing into the load.

From equation (8.12),

$$V_{An} = \frac{m_A}{2\sqrt{2}} V_{in} = \frac{0.6 \cdot 550}{2\sqrt{2}} = 116.6\ \text{V}$$

The per phase load impedance is

$$\overline{Z}_A = 8 + j(2\pi 90) \cdot 2.3 \times 10^{-3} = 8.11\ \Omega \underline{/9.2^\circ}$$

Assuming that V_{An} is at an angle of zero,

$$\overline{I}_A = \frac{V_{An} \underline{/0^\circ}}{\overline{Z}_A} = 14.21\text{A} \underline{/-9.2^\circ}$$

The three phase real and reactive power flow due to the fundamental current and voltage are then

$$S = 3 \cdot V_{An} \cdot I_A = 3 \cdot 116.6 \cdot 14.21 = 5038\ \text{VA}$$
$$P = S \cdot \cos(\delta_A - \phi_A) = 5038 \cdot \cos(0^\circ - (-9.2^\circ)) = 4973\ \text{W}$$
$$Q = S \cdot \sin(\delta_A - \phi_A) = 5038 \cdot \sin(0^\circ - (-9.2^\circ)) = 809\ \text{VARs}$$

The converter input power can be calculated from the output power and converter efficiency. The input current can then be calculated from the input power.

8.3 Converter connected synchronous machines

The per phase equivalent circuit for a converter connected synchronous machine is shown in figure 8.13. While it shows only one phase of the machine, keep in mind that it is a three phase unit, as shown in figure 8.12. Also note that the electrical quantities on the left side of this diagram are DC. The notation on these has been changed in figure 8.13 to emphasize this. The electrical quantities on the right side are alternating, with values expressed as phasor quantities, as is appropriate for sinusoidal steady state operation.

Also note that the synchronous machine model is shown somewhat differently, to accommodate the variable frequency capability of the converter. In particular, the inductive reactance is shown changing with frequency, and the internal machine voltage is shown as being proportional to both frequency and field current. Recall that permanent magnet machines are modeled as having a fixed field current.

Figure 8.13 shows the system in motor convention, where the reference direction for power and current flow are into the machine. The phase voltage $\overline{V}_{An} = V_{An}\,\underline{/\delta_A}$ is the RMS fundamental voltage from machine A phase to the machine neutral point. The amplitude modulation ratio m_A can then be found from equations (8.12) or (8.14), depending on the modulation strategy used in the converter. The phase current $\overline{I}_A = I_A\,\underline{/\phi_A}$ is the RMS fundamental current flowing into the A phase terminal. The B and C phase voltages and currents are equal in magnitude and phase shifted by 120° and 240° respectively.

Figure 8.14 shows a phasor diagram for the case where phase current I_A is in phase with E_A. The motor terminal voltage can be found from the phasor diagram shown in figure 8.14.

Remember that motor shaft speed is directly proportional to applied electrical frequency ω_e. With $\overline{I}_A$ constant and in phase with $\overline{E}_{Af}$, the phasor diagram will expand proportionally as ω_e increases—apart from the small voltage drop $R_A\overline{I}_A$.

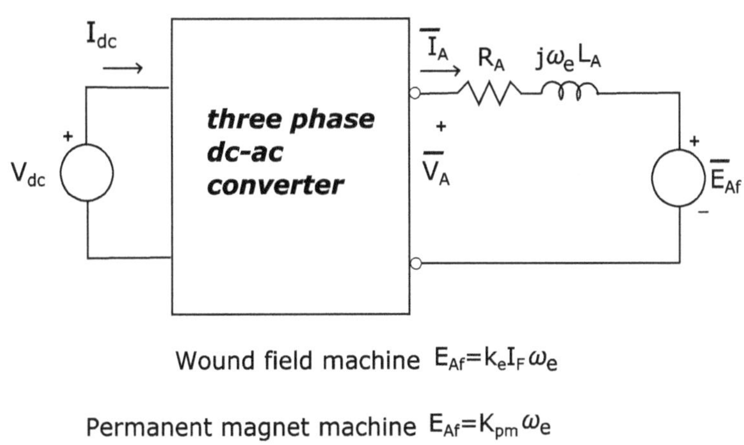

Figure 8.13. Per phase equivalent circuit of round rotor synchronous machine connected to a power converter. Motor convention.

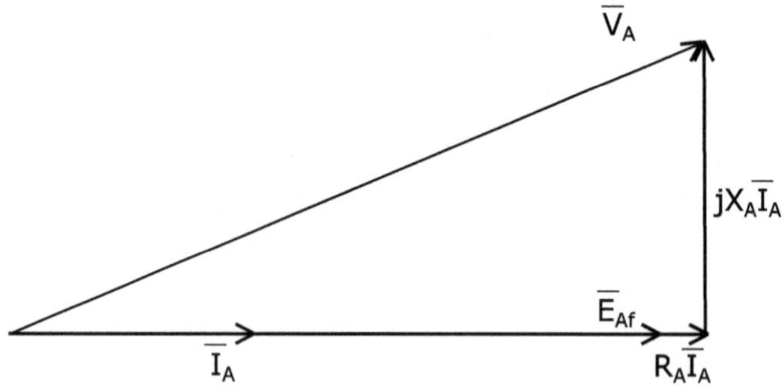

Figure 8.14. Phasor diagram for synchronous motor.

The three phase real power flowing from the converter into the machine is

$$P_s = 3V_A I_A \cos(\delta_A - \phi_A) \tag{8.15}$$

Similarly, the fundamental reactive power flowing into the machine is

$$Q_s = 3V_A I_A \sin(\delta_A - \phi_A) \tag{8.16}$$

If the converter efficiency is η, then the DC power flowing into the converter is

$$P_{dc} = \frac{1}{\eta} P_S \tag{8.17}$$

In figure 8.13, DC power is

$$P_{dc} = V_{dc} I_{dc} \tag{8.18}$$

With P_{dc} and V_{dc} known, the DC input current can be found. In cases where the DC source is not ideal, the DC side model can be expanded to represent the actual source network.

The motor converted power for this round rotor machine is

$$P_{conv} = 3E_A I_A \cos(\delta_{EA} - \phi_A) \tag{8.19}$$

The developed shaft torque is then

$$T_e = \frac{P_{conv}}{\omega_r} = \frac{\text{poles}}{2} \frac{P_{conv}}{\omega_e} \tag{8.20}$$

In this motor, the motor must be operated within its current and voltage ratings. The motor voltage will vary with frequency, and the motor current will vary with torque. The motor will be able to operate with rated torque for shaft speeds from standstill up to rated shaft speed. At rated speed, the motor terminal voltage will reach its rated value. Above rated speed, a component of stator current can be used to demagnetize the motor and keep the voltage at its rated value. This will result in a reduction of the torque production capability of the motor at speeds above rated.

Example 8.4. A three phase PWM bridge is feeding a balanced three phase RLE load. The per phase equivalent of the circuit is shown in figure 8.13. The DC input voltage is 550 V. The inverter is operating in sinusoidal PWM control, with $m_A = 0.55$, and $f_1 = 45\text{Hz}$.

(a) Determine the magnitude of the line to neutral output voltage

$$V_{An} = 0.354 m_A V_{in} = 0.354 \cdot 0.55 \cdot 550 = 107.1\ \text{V}$$

The line to line voltage magnitude is then

$$V_{AB} = \sqrt{3} \cdot 107.1 = 185.5\ \text{V}$$

(b) If $R_A = 0.12\Omega$, $L_A = 0.5\text{mH}$, and the line current is $I_A = 82\text{A}$,with the current in phase with the inverter output voltage V_{An}. Find the real and reactive power flow to the load, and the internal voltage $\overline{E}_{Af}$.

As the line current is in phase with the line to neutral voltage, there is no reactive power flow from the inverter. The three phase apparent power is

$$S_I = 3 \cdot V_{An} \cdot I_A = 26.35\ \text{kVA}$$

Because there is no reactive power flow, the real power equals the apparent power,

$$P_I = 26.35\ \text{kW}$$

The inductive reactance of the load is

$$X_A = 2\pi f_1 L_A = 0.141\ \Omega$$

The internal voltage is

$$\overline{E}_{Af} = \overline{V}_{An} - (R_A + jX_A) \cdot \overline{I}_A$$
$$\overline{E}_{Af} = 107.1\ \underline{/0^\circ} - (0.12 + j0.141\ \Omega) \cdot 82\ \underline{/0^\circ}$$
$$\overline{E}_{Af} = 98.0\ \underline{/-6.8^\circ}\text{V}$$

Example 8.5. A three phase inverter is supplying a round rotor four pole permanent magnet synchronous motor. The per phase equivalent circuit of the inverter/motor system is shown in figure 8.13. The permanent magnet synchronous motor constant is $K_{PM} = 0.45$.

(a) The motor is operating at 750 rpm. Determine the frequency of the inverter.

$$f_e = 750\ \text{rpm} \cdot \frac{(4/2)}{60} = 25\ \text{Hz}$$

(b) Determine the magnitude of the line to neutral internal voltage E_{Af} generator voltage.

$$E_{Af} = K_{PM}\omega_e$$

The electrical frequency in radians per second is

$$\omega_e = 2\pi \cdot 25 = 157 \text{ rad/sec}$$

So the internal voltage is

$$E_{Af} = 0.45 \cdot 157 = 70.65 \text{ V}$$

(c) The current I_A has magnitude of 80 A. This current leads the internal voltage E_{Af} by 30°. Assume that E_{Af} is at an angle of 0° and find the terminal voltage of the motor. The motor resistance is 0.16 ohms, and the motor inductance is 5.8 millihenries. Find the line to neutral terminal voltage of the motor.

$$\overline{Z}_s = R_a + j\omega_e L_s = 0.16 + j7.85 \text{ ohms}$$

The line to neutral terminal voltage is then

$$\overline{V}_{An} = \overline{E}_{Af} + \overline{Z}_s \cdot \overline{I}_A = 70.65\text{V} \underline{/0^\circ} + (0.16 + j7.85) \cdot 80\text{A} \underline{/30^\circ}$$

or

$$\overline{V}_{An} = 83.0 \text{ V} \underline{/57^\circ}$$

(d) Find the real and reactive power flowing into the motor.

$$P_s = 3 \cdot 83.0 \cdot 80 \cdot \cos(57^\circ - 30^\circ) = 17\,750 \text{ W}$$
$$Q_s = 3 \cdot 83.0 \cdot 80 \cdot \sin(57^\circ - 30^\circ) = 9043 \text{ VARs}$$

(e) The inverter input is $V_{dc} = 600$ V. Find the amplitude modulation ratio for this operating point.

From equation (8.12), $V_{An} = 0.354 m_A V_{in}$ or $83.0\text{V} = 0.354 m_A 600\text{V}$

Solving for m_A, $m_A = \frac{83.0}{0.354 \cdot 600} = 0.391$

(f) The motor has friction and windage loss of 400 W. Determine the shaft power and torque for the motor.

The stator loss is: $P_{sLoss} = 3 \cdot 80 \cdot 0.16 = 38.4$ W

The output (shaft) power is $P_{out} = P_s - P_{sLoss} - P_{f,\,w\,Loss} = 17250 - 38.4 - 400 = 16811$ W

The shaft speed in mechanical radians per second is

$$\omega_r = \frac{\omega_e}{\text{poles}/2} = 78.54 \text{ rad/sec}$$

The output torque is then $T_{shaft} = \frac{P_{out}}{\omega_r} = \frac{16811 \text{ W}}{78.54 \text{ rad/sec}} = 214$ N m

8.4 The ideal DC drive

In order to understand AC drive system performance, it is instructive to first discuss an idealized version of separately excited DC drives. This type of DC drive is fed by two independent voltage sources, one for the armature and the other for the field. The circuit diagram for this drive is shown in figure 8.15.

While both armature and field windings will have inductance, there will be no voltage drop across the inductance in the steady state. The resulting equation for the armature winding is

$$\boldsymbol{V}_{\mathbf{A}} = \boldsymbol{R}_{\mathbf{A}}\boldsymbol{I}_{\mathbf{A}} + \boldsymbol{E}_{\mathbf{fd}} \tag{8.21}$$

The field winding equation is

$$\boldsymbol{V}_{\mathbf{f}} = \boldsymbol{R}_{\mathbf{f}}\boldsymbol{I}_{\mathbf{f}} \tag{8.22}$$

The armature voltage induced by the field current is

$$E_{\mathrm{fd}} = K_{\mathrm{fd}}I_{\mathrm{f}}\omega_{\mathrm{r}} \tag{8.23}$$

The converted power is

$$P_{\mathrm{conv}} = E_{\mathrm{fd}}I_{\mathrm{A}} = \omega_{\mathrm{r}}T_{\mathrm{e}} \tag{8.24}$$

From this equation, the developed torque in newton meters is then

$$T_{\mathrm{e}} = K_{\mathrm{fd}}I_{\mathrm{f}}I_{\mathrm{A}} \tag{8.25}$$

The DC motor will have a rated point. The significant ratings from a speed control point of view are given in table 8.4. When the motor operates at rated speed with rated output power, it will be operating with rated armature voltage and current and rated field voltage. These are the maximum allowable steady state values for these voltages and currents, plus output torque and power. However, the rated speed is not the maximum speed, and the motor can operate safely at significantly higher speeds.

The DC drive has two distinct operating regions—constant torque and constant power.

Constant torque operation: The DC drive operates in the constant torque region when the shaft speed is less than rated speed. In this region, the field voltage is maintained at its rated value. The internal voltage E_{fd} is directly proportional speed. The armature terminal voltage is the internal voltage plus the voltage drop across the armature resistance, $R_{\mathrm{A}}I_{\mathrm{A}}$. As a result, the armature terminal voltage will be less than rated voltage in the constant torque region. This resistive voltage drop will be small relative to the internal induced voltage. The armature current must be no greater than rated current, and is the limiting factor in constant torque operation.

Therefore, at rated armature current, equation (8.25) states that the developed torque capability of the motor is constant and independent of the speed. The drive can actually operate at any developed torque below this rated value. Equation (8.24) shows that the output shaft power will be proportional to speed when operating in this constant torque region.

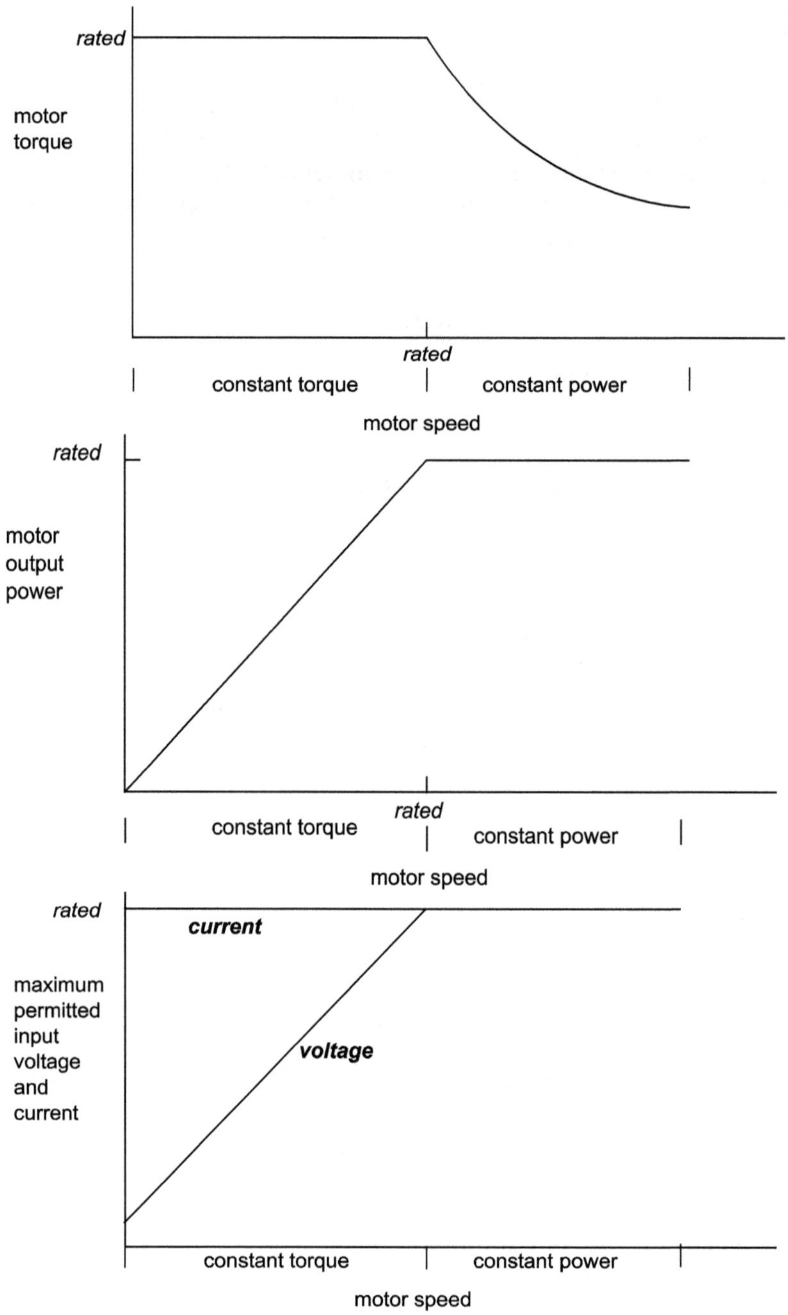

Figure 8.15. Motor operating limits showing the constant torque and constant power operating regions.

Table 8.4. DC motor rated quantities and their interpretation.

Rated quantity	Note
Speed	Operation above rated speed is allowed
Output torque	Maximum allowable value
Output power	Maximum allowable value
Armature voltage	Maximum allowable value
Field voltage	Maximum allowable value
Armature current	Maximum allowable value

In order to operate in this region, the armature must be supplied with a variable voltage source. For this region, constant torque operation is sometimes referred to as the armature control region.

Constant power operation: When the shaft speed goes above rated speed, the field current I_f must be reduced in order to keep the internal voltage E_{fd} and the armature terminal voltage at or below their maximum allowable values. Equation (8.24) shows that the converted power will be constant when E_{fd} is constant and the armature current is constant. Equation (8.23) shows that in order to keep E_{fd} constant, the field current must be inversely proportional to the speed. As a result, equation (8.25) shows that the allowable converted torque will fall off in inverse proportion to the shaft speed. For these reasons, motor operation above rated speed is referred to as the constant power region, or the field weakening region.

The resulting plots of torque, power, armature voltage and armature current are shown in figure 8.15. Note that these curves are capability curves, in that the motor can be operated safely in the region below the curve. The capability differences between the constant torque and the constant power regions are clear in these plots, and the motor capability must be carefully matched with the load requirements in a specific application to get the best overall drive system.

As would be expected, there is an upper limit on the motor speed. This upper speed is set by factors other than voltage and current ratings, such as mechanical considerations for the bearings, winding stresses, and commutation limits.

While the DC machine discussed in this section is conceptually straightforward, the motor itself is more complex, more costly, and less reliable than comparable AC machines. With the advent of low cost, highly flexible inverters, variable speed AC drives are supplanting DC drives in most applications.

8.5 Variable speed round rotor permanent magnet synchronous motor drives

Three phase permanent magnet synchronous motors are one of the prominent variable speed drive technologies today. There are two basic designs for the permanent magnets that set up the fields on the rotors of these machines

- Surface mounted magnets.
- Interior permanent magnets.

These designs are discussed in chapter 5.

The following discussion is for surface mount designs, with no saliency. The per phase diagram of figure 8.13 is then the model used for this analysis.

Permanent magnet synchronous machines have many of the same ratings/limitations as the DC machine, as described in table 8.4. The primary exception is that the field flux level is set by the permanent magnets, and therefore is constant. Also, the armature has AC voltages and currents rather than DC quantities, but the nature of the armature limits are the same as in the table.

Constant torque region: When the motor is operating below rated speed/frequency, the internal voltage $E_{Af} = K_{pm}\omega_e$ is proportional to speed/frequency, and is operating below its maximum permissible value. Remember that we have stated E_{Af} in terms of electrical frequency ω_e rather than shaft speed ω_r in this equation, and that these two values are related by the ratio poles/2.

As motor shaft speed increases, both ω_e and E_{Af} will increase in direct proportion to shaft speed ω_r. The motor terminal voltage is the internal voltage plus the drop across the armature impedance $R_A + j\omega_e L_A$, and will be somewhat larger than the internal voltage. The limit for the armature current will again be the rated current, just as in the DC machine.

One constant torque region control approach is to keep the armature current in phase with the machine internal voltage in this operating region. We will choose both E_{Af} and I_A to be at 0°. If the armature resistance R_A is neglected, the terminal voltage will then be

$$\overline{V_A} = V_A \underline{/\delta} = j\omega_e L_A I_A + E_{Af} = j\omega_e L_A I_A + K_{pm}\omega_e \tag{8.26}$$

For a given value of armature current I_A, the magnitude of V_A will be proportional to speed, and the voltage angle δ will be constant. Note that there is a correlation between this phasor equation and the space vector relationships presented in chapter 5. In terms of space vectors, the E_{Af} space vector is aligned on the Q axis, and the field flux vector is on the D axis as shown in figure 5.15.

The power entering the motor will be

$$P_{conv} = 3V_A I_A \cos(\delta) \tag{8.27}$$

With armature resistance neglected, the input power will equal the converted power. From equation (8.26), $V_A\cos(\delta) = E_{Af}$, so

$$P_{conv} = 3E_{Af} I_T \tag{8.28}$$

The developed torque will then be

$$T_e = \frac{P_{conv}}{\omega_r} = \frac{3 \cdot \text{poles}}{2} K_{pm} I_T \tag{8.29}$$

Note that in the final forms of these two equations, the armature current I_A is replaced by the current I_T. I_T is referred to as the torque producing current. It is the component of armature current that is aligned with the internal voltage E_{Af}. When I_A is aligned with E_{Af}, all of this current goes to produce torque in this control mode, and the two quantities are equal.

In summary, the pm synchronous motor acts very similarly to the ideal DC motor in the constant torque region. The developed torque T_e is constant for a given armature current, and the converted power increases in proportion to speed. Rated torque is produced by rated current, and the machine will reach rated terminal voltage at rated speed and rated torque.

Constant power region: Above rated speed, the internal voltage rises above the rated terminal voltage. In this case, field weakening is needed for this machine, as it is for the ideal DC machine. In this machine, however, the field flux is provided by permanent magnets, and has a constant magnitude. From chapter 5, the field flux lies on the direct (D) axis. If a component of stator current is injected on that same axis, but in the negative direction, it will create D axis flux in the opposite direction as the permanent magnets, and will therefore reduce the resultant D axis flux level.

When the motor shaft speed is greater than rated speed, the net internal voltage induced by the magnetic flux due to the permanent magnets and the demagnetizing stator current should equal the voltage induced by the permanent magnets at rated speed. With stator resistance neglected, the D axis equation becomes:

$$E_{\mathrm{AfR}} = K_{\mathrm{pm}}\omega_{\mathrm{eR}} = K_{\mathrm{pm}}\omega_{\mathrm{e}} - \omega_{\mathrm{e}}L_{\mathrm{A}}I_{\mathrm{F}} \quad \text{for } \omega_{\mathrm{e}} > \omega_{\mathrm{eR}} \tag{8.30}$$

Here, ω_{eR} is the rated electrical frequency of the motor, and E_{AfR} is the internal machine voltage at the rated frequency operating point. I_{F} is termed the flux producing current, and lies on the negative D axis. As a result, positive values of I_{F} will reduce the D axis flux and hence the internal machine voltage.

The torque producing current I_{T}, which is on the quadrature (Q) axis, must also flow in these same stator windings. The total stator current must stay below rated current, so that the maximum allowable level of torque producing current is:

$$I_{\mathrm{Tmax}} = \sqrt{I_{\mathrm{Arated}}^2 - I_{\mathrm{F}}^2} \tag{8.31}$$

From figure 8.13, the motor terminal voltage is (neglecting stator resistance):

$$\overline{V}_{\mathrm{A}} = j\omega_{\mathrm{e}}L_{\mathrm{A}}\overline{I}_{\mathrm{A}} + E_{\mathrm{Af}} \tag{8.32}$$

The phasor diagram for this equation is shown in figure 8.16.

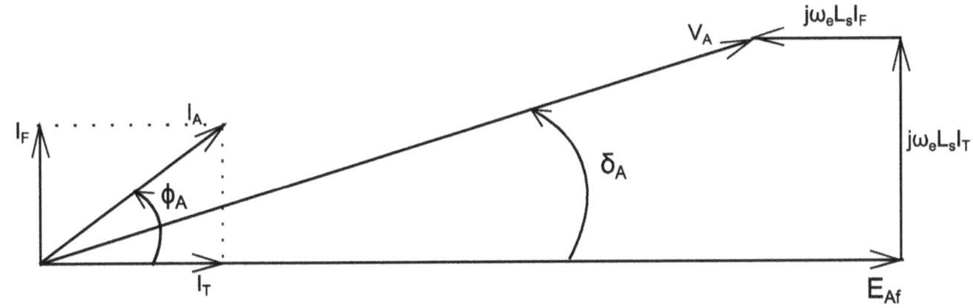

Figure 8.16. Phasor diagram for equation (8.32), showing the torque producing and flux producing components of the stator current.

The motor terminal voltage magnitude must be kept below rated voltage,

$$V_A \leq V_{Arated} \tag{8.33}$$

The power entering the motor will then be

$$P_{in} = 3V_A I_A \cos(\delta_A - \phi_A) \tag{8.34}$$

Here, E_{Af} lies on the reference Q axis, δ_A is the angle that V_A leads the Q axis, and ϕ_A is the angle that I_A leads the Q axis. Note that, with the stator resistance neglected, the converted power will equal the input power.

Through trigonometric substitution,

$$P_{conv} = 3\omega_e K_{pm} I_T \tag{8.35}$$

While this is the same equation as for the constant torque region, the value of I_T must be decreased by the need to keep the stator current less than rated and the stator voltage less than rated. While this decrease in I_T does offset the increase in ω_e, this is not necessarily an exact cancellation. As a result, the maximum power that can be delivered by this motor does have some variation as the speed rises above rated speed. These and other factors (including armature resistance and motor magnetic field and rotational losses) make a complex optimization problem as to the steady state and short term limits on motor power and torque at a given speed, and on the best operating point for the motor within the power limitation.

There are several factors that limit the motor top speed. There are a number of mechanical stress factors, including stresses on the stator windings and rotor assembly, including the permanent magnets. On the electrical side, limits include the point where the flux producing current becomes equal to the rated stator current, and the point at which the stator voltage cannot be maintained at or below rated voltage.

It is worth mentioning that, in the case of salient pole rotors the inductance is different for the torque producing current and the flux producing current. If you adjust figure 8.16 accordingly and conduct this same analysis, you will find the both the converted power of developed torque equations are different for the salient pole machine.

Example 8.6. For the motor of example 8.5, neglect the motor resistance. The current magnitude is 80 A, and it is in phase with the internal voltage E_{Af}.

a. Plot E_{Af} and V_{An} versus frequency, for frequencies up to 60 Hz.

Assume that E_{Af} is at an angle of 0°. The internal motor voltage is

$$E_{Af} = K_{PM}\omega_e = 0.45 \cdot 2\pi \cdot f_e = 2.83 f_e$$

The voltage drop across the stator inductance is

$$V_X = j\omega_e L_S I_A = j2\pi \cdot 0.005\text{ H} \cdot 80\text{ A} \cdot f_e = j2.513 f_e$$

The stator terminal voltage is then

$$\overline{V}_{An} = E_{Af} + V_X = (2.83 + j2.513)f_e$$

The magnitude of the terminal voltage is then

$$V_{An} = 3.78f_e$$

The plot of E_{Af} and V_{An} versus input frequency is shown below.

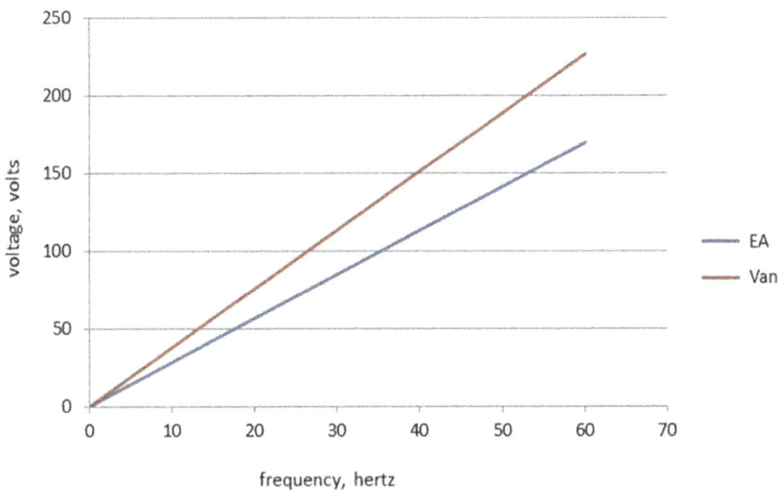

b. The DC voltage at the inverter input terminals is 600 V, and the inverter is modulated using SPWM. Find the amplitude modulation ratio as a function of frequency.

From equation (8.12), $V_{An} = 0.354m_A V_{in}$ or $83.0\text{ V} = 0.354m_A 600\text{ V}$

Therefore, $m_A = \frac{V_{An}}{0.354 \cdot 600} = \frac{3.78 \cdot f_e}{212.4} = 0.0178f_e$.

Note that the maximum practical value of m_A for a sinusoidal PWM inverter is 1.0. For these values of inverter input voltage and stator current, this will happen at a frequency slightly below 60 Hz. For a range of higher frequencies, the motor can be operated at frequencies above 60 Hz, with a constant value of m_A and hence V_{An}. The machine can be operated in this range, although with a reduced torque output capability.

c. What is the torque developed by the motor as a function of frequency, for frequencies up to 60 Hz? Friction and windage is neglected in determining the developed torque.

The converted power is the product of internal voltage and line current. As the current is in phase with the internal voltage in this example,

$$P_{conv} = 3E_{Af}I_A = 3 \cdot 2.83f_e \cdot 80\text{ A} = 679.2f_e$$

The converted torque is

$$T_e = \frac{P_{conv}}{\omega_r} = \frac{679.2}{\pi} = 216.2\text{ N m}$$

The torque is constant under these conditions. This requires that the inverter amplitude modulation ratio be controlled to keep the stator current at 80 A and in phase with the internal voltage as motor speed changes.

Example 8.7. The permanent magnet motor of example 8.6 is rated at 60 Hz, 4 poles, 227 V line to neutral, and 80 A. At the rated frequency, $E_{Af} = 170$ V. Find an acceptable operating point when the frequency is 90 Hz. What are the voltage, current, speed, power and torque production at this point?

At 90 Hz,$\omega_e = 2\pi 90 = 565.5$ rad/sec. Then

$$E_{Af} = 0.45\omega_e = 255\text{ V}$$

From equation (8.30), $170\text{V} = 255\text{V} - \omega_e L_A I_F$ so an initial value for flux producing current is

$$I_F = \frac{255 - 170}{565.5 \cdot 0.005} = 30.0\text{ A}$$

With this current, the maximum torque producing current would be

$$I_{Tmax} = \sqrt{80^2 - 30^2} = 74.2\text{ A}$$

With this value of torque and flux producing currents, the terminal voltage would be

$$\overline{V}_A = 255\text{V} - 85\text{V} + j565.5 \cdot 0.005\text{H} \cdot 74.2\text{ A} = 269.7\text{ V}\,\underline{/51.0^\circ}$$

This voltage exceeds the motor rating. If the torque producing current is reduced below its limit in proportion to the shaft speed,

$$I_T = 80\text{ A} \cdot \frac{60\text{ Hz}}{90\text{ Hz}} = 53.3\text{ A}$$

With this current, the motor terminal voltage is

$$\overline{V}_A = 255\text{ V} - 85\text{ V} + j565.5 \cdot 0.005\text{ H} \cdot 53.3\text{ A} = 227\text{ V}\,\underline{/41.6^\circ}$$

This operating point is within the motor's voltage and current ratings. The motor current magnitude is

$$I_s = \sqrt{30^2 + 53.3^2} = 61.2 \text{ A}$$

Motor power and torque are

$$P_{\text{in}} = 40.17 \text{ kW}$$
$$T_e = 144 \; N \; m$$

While this operating point is within the motor ratings, the current is less than rated. For this particular motor, there is therefore room to further adjust the torque and flux producing currents to increase the current magnitude while keeping the stator voltage within its rating.

8.6 Variable speed induction motor drives

Figure 8.17 shows the per phase equivalent circuit of a converter fed three phase induction machine. The figure shows current and power reference directions chosen in the motor convention. As with the synchronous machine, the induction motor per phase equivalent circuit is chosen to reflect that the converter frequency ω_1 is variable.

Also note that the rotor resistance element in the equivalent circuit is represented as

$$\frac{R_r\omega_1}{s\omega_1} = \frac{R_r\omega_1}{\omega_2} \tag{8.36}$$

This representation is convenient in variable speed induction motor drives, where the slip frequency $\omega_2 = s\omega_1$ is often held constant or nearly constant. The rotor

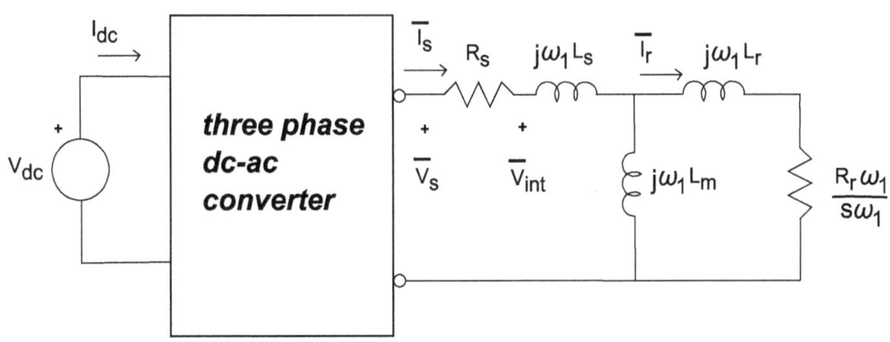

Figure 8.17. Per phase equivalent circuit of a squirrel cage induction machine connected to a power converter. Motor convention.

electrical frequency is often called the slip frequency, particularly in variable speed induction motor drives. It is important to note that, if $s\omega_1$ is held constant, then the apparent rotor resistance element described in equation (8.21) is proportional to converter frequency ω_1. The inductive reactances of the motor are similarly proportional to converter frequency. Therefore, apart from the stator resistance R_1, the motor impedance varies in proportion with the converter frequency. If the applied stator voltage magnitude is also varied in direct proportion with the frequency, the resulting current flow and developed torque will be approximately constant.

A given induction machine will have a voltage, current, frequency, output power, and speed rating. The voltage, current and output power ratings reflect the maximum allowable steady state quantities for these parameters. The frequency and speed ratings, however, can be exceeded. The rated frequency does again form the boundary between two quite different operating modes, however.

Constant torque operation: The motor operating region below rated frequency/speed is referred to as the constant torque region. In the constant torque region, motor performance is limited by the machine's current and torque ratings. Both machine voltage and output power will be less than their rated values.

Consider an induction motor operating at its rated point—rated speed, frequency, slip frequency, voltage, current, developed torque and power.

Now consider the situation when the stator frequency ω_1 is cut in half while the slip frequency ω_2 remains at its rated value $\omega_{2(\mathrm{R})}$. In figure 8.15, all motor impedances will be cut in half, with the exception of R_1. For the motor to draw rated current from the converter, the internal voltage V_{int} would need to be cut in half as well. Since the voltage drop across R_1 is not changed by this changed frequency, the applied stator voltage V_1 would need to be slightly larger than half of its rated value.

Extending this to the full range of applied frequencies less than rated stator frequency, it can be seen that rated torque can be achieved by maintaining operation at rated slip frequency with a voltage approximately proportional to the applied frequency. Operation at rated current and rated slip frequency will supply rated developed torque. This will be the maximum torque that the motor can supply at a given operating speed, and the output power will be the product of this torque and the motor shaft speed. The maximum output power in this operating region will therefore be proportional to shaft speed. The motor can, of course, drive mechanical loads at less than rated torque. This is generally done by reducing the operating slip frequency of the motor. Alternate strategies have been proposed that have a goal of minimizing losses.

Constant power operation: As indicated earlier, most motor designs allow for operation above rated speeds. As the voltage cannot exceed the rated voltage, however, the motor cannot draw the rated current when operating above the rated speed if it is operating at rated slip frequency. If the slip frequency is allowed to increase, however, the motor can still draw the rated current as the decrease in apparent rotor resistance would compensate the increase in impedance of the inductive reactances of the machine. As developed torque is inversely proportional to slip frequency, this torque will decrease as the motor speed increases. However, the product of torque and speed will remain approximately constant—and this is the converted power of the motor. Therefore, the induction machine is capable of delivering rated power or near rated power in this operating region. This operating region can be extended up to the point where the developed torque reaches the motor breakdown torque on the motor torque–speed curve. This is the upper limit of the constant power region of the induction machine. Operation above this point can be possible, but at reduced power output capability.

Again, rated voltage and rated power define the maximum operating conditions in the constant power region. The machine can operate below maximum power, through adjusting voltage and/or slip frequency with a resulting reduction in line current. As a result, the induction motor performance curves are similar to the ideal DC machine performance curves of figure 8.15, up to the point where the slip reaches the motors maximum torque point.

Example 8.8. A three phase 4 pole induction motor is fed from a PWM inverter. The per phase equivalent circuit of this drive is shown in figure 8.15. The motor electrical constants are:

R_1	L_1	L_m	L'_2	R'_2
0.02 ohms	0.160 mH	9.28 mH	0.160 mH	0.015 ohms

Find the motor current, input power and air gap torque when operating at a terminal voltage of 80 v line to neutral, a frequency of 40 Hz, and a slip frequency of 1.5 Hz.

The stator frequency in radians per second is $\omega_1 = 2\pi 40 = 251.3\frac{\text{rad}}{\text{sec}}$. The induction motor reactances are then

$$X_1 = X'_2 = 251.3 \cdot 0.000\,160 \text{ H} = 0.040\ \Omega$$
$$X_M = 251.3 \cdot 0.00\,928 \text{ H} = 02.332\ \Omega$$

The slip $= \frac{1.5}{40} = 0.0375$, and

$$\frac{R_2'}{s} = 0.40$$

The per phase equivalent circuit for this case is shown below.

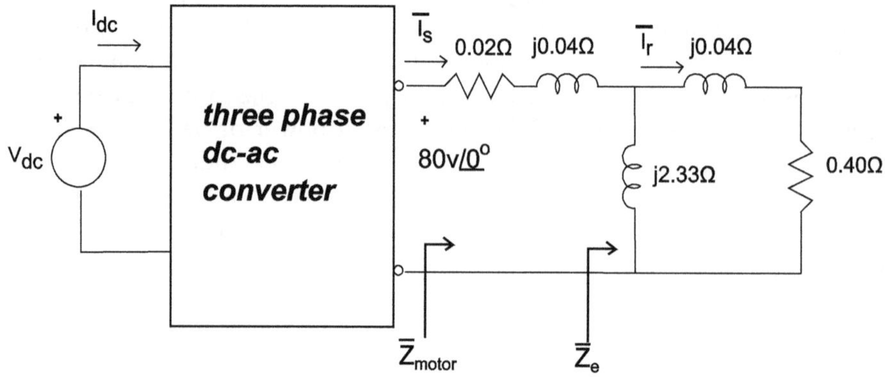

The parallel combination of the rotor and magnetizing impedances is

$$\overline{Z}_e = \frac{j2.33(0.40 + j0.04)}{j2.33 + (0.40 + j0.04)} = 0.3760 + j0.1028\ \Omega$$

The motor per phase impedance is then

$$\overline{Z}_{\text{motor}} = 0.02 + j0.04 + \overline{Z}_e = 0.3959 + j0.1428\ \Omega$$

The stator current is

$$\overline{I}_s = \frac{80\text{V}\,\underline{/0^\circ}}{\overline{Z}_{\text{motor}}} = 190.1\ \text{A}\,\underline{/-19.8^\circ}$$

The input power of the motor is

$$P_S = 3(80\ \text{V})(190.2)\cos(0^\circ - (-19.8^\circ)) = 42.92\ \text{kW}$$

The stator power loss is

$$P_{\text{stator loss}} = 3(190.2)^2(0.02) = 2.17\ \text{kW}$$

The air gap power is then

$$P_{\text{air gap}} = P_S - P_{\text{stator loss}} = 40.75\ \text{kW}$$

The rotor current is found from the current divider,

$$\overline{I}_R = \frac{jX_M}{jX_M + \overline{Z}_R}\overline{I}_s = 184.3\ \text{A}\,\underline{/-10.3^\circ}$$

From equation (7.38), the developed torque is

$$T_e = \left(\frac{3 \cdot 4}{2 \cdot 251.3}\right)184.3^2(0.4) = 216.2 \text{ N m}$$

Note that the developed torque can also be found from the air gap power divided by the synchronous speed in mechanical radians per second ($2\omega_1$/poles).

8.7 AC motor drive performance and control

Variable speed drives have the capability to operate over a wide range of speed and torque. In a given application with a given load, a controller is needed whose task is to set the drive operating point at a given set point. In general, the drive could be controlling on motor torque or speed, regulating the pressure or flow rate of a pump, or perhaps providing torque to drive a vehicle. In turn, the set point will be changed to accommodate changing needs of the larger system—much as a thermostat command is changed when someone in the room wants a cooler temperature.

A good example would be an induction motor drive in an electric vehicle. In this case, the vehicle operator sets the command with the throttle (which may still called the gas pedal even in EVs). The throttle command is essentially a torque command. The driver of the vehicle selects a torque command for initial acceleration from standstill, and then adjusts this setting as the vehicle approaches the operating speed. Essentially, the human driver is in the control loop, setting the torque command based on the vehicle speed and the driving criteria. As you are no doubt aware, cruise control is available to maintain a given speed in many vehicles. However, cruise control may not work well during vehicle acceleration, including hill climbing, where it can lead to excessive up/down shifting in some vehicles.

Figure 8.18 shows a conceptual diagram for one type of field oriented AC motor controller. In field oriented control, the AC motor currents are transformed into two components—the direct or torque producing current i_Q, and the quadrature or flux producing current i_D. The direct and quadrature currents are constant when the motor is at a steady operating point, but are slowly varying quantities during motor acceleration. As a result, voltage and current variables are generally represented by lower case variables in these diagrams.

A key to field oriented control is knowing the position of the machine flux waves. In the PM synchronous machines, the position of the Q axis is needed for the motor control. As the Q axis is at a fixed position on the rotor, a rotor position sensor can be used to measure the rotor angle. When this position is known, the phase currents can be resolved into their i_Q and i_D components. These quantities are then compared with the controller torque and flux commands, and desired Q and D voltage commands are developed. These in turn are converted back into their abc components, and the phase voltage commands are then sent to the PWM controller in the inverter. Some drives use rotor position estimators to avoid the position sensor, and these estimators have suitable accuracy for many applications.

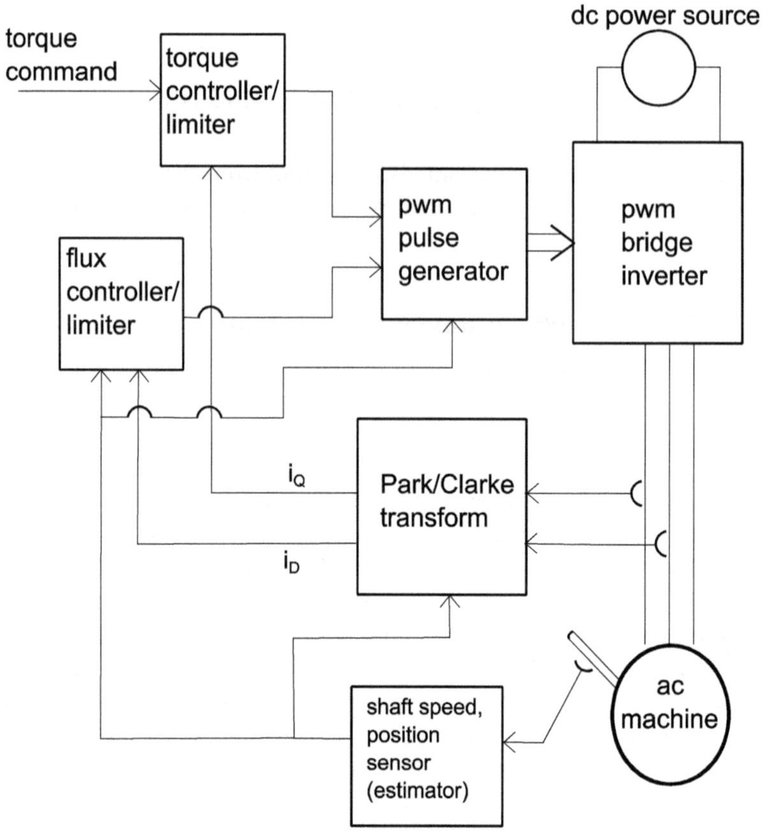

Figure 8.18. Conceptual diagram, field oriented AC motor control.

Induction motor controllers can be more involved as the flux wave does not have a fixed reference point on the rotor. These controllers start by estimating the Q axis position, and there are a variety of methods controllers use to do this. Once the Q axis location is determined, the control operates in a similar fashion to the PM motor controller, controlling on the torque and flux producing currents, and developing voltage commands that are sent to the inverter.

At present, there are many different approaches to implementing AC drives controls, and the block diagram of figure 8.18 shows the basic components of these controllers at a conceptual level.

Questions

Search the web for a specification sheet for a power

1. IGBT rated at 1200 V and approximately 200 A. From this document,
    a. What is the part number of the device, and the manufacturer?
    b. What is the on state (collector–emitter saturation) voltage?
    c. What is the gate threshold voltage?
    d. What is the maximum junction temperature?

e. What are the rise and the fall times for this device? If you assume that the minimum switching period is at least 10 times the sum of these two values, what is the minimum switching period? What then is the maximum switching frequency?
f. Power IGBTs are sometimes packaged as single devices, sometimes as a full power pole, and sometimes as a full three phase converter. How is the device that you found packaged?

Problems

1. The three phase PWM converter shown below is operating with space vector PWM control. The input voltage V_{in} = 250 V DC, the amplitude modulation ratio is $m_A = 0.26$, $R_A = 5$ ohms, $L_A = 0.8$millihenries, and the output frequency $f_1 = 100$ Hz. Neglect converter losses.
   a. Find the A phase current magnitude, and its phase angle relative to the A phase to neutral voltage.
   b. Find the real and reactive power flow to the load due to the 100 Hz quantities.
   c. Find the DC input power and input current I_{in}.

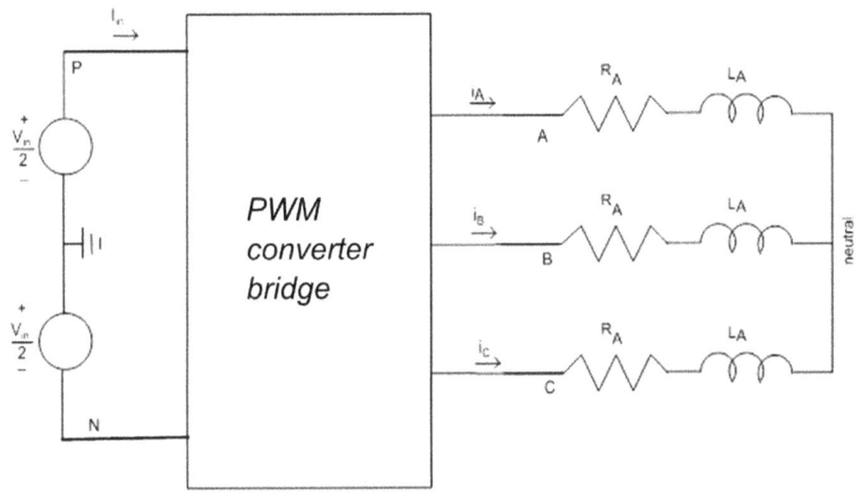

2. This same converter and load is operating under balanced conditions with an output frequency of 40 Hz and an A phase current of 15 A rms. Find
   a. The amplitude modulation ratio m_a.
   b. The real and reactive power output of the bridge, based solely on the fundamental frequency voltage and current (remember that m_a must be between 0 and 1 for a valid solution).
3. A converter bridge is operating with a fundamental output frequency of 25 Hz, and a switching frequency modulation ratio of 33.
   a. What is the converter switching frequency?

b. The converter output voltage has the following significant harmonic components:

Harmonic (n)	RMS L-N harmonic voltage (V_n)
Fundamental	120 V
33rd	98.2 V
65th	37.7 V
67th	37.7 V

Calculate the total RMS value of these voltages $\left(V_{\text{total rms}} = \sqrt{\sum_n V_n^2}\right)$

c. The converter is loaded with a balanced three phase RL load as shown in problem 8.1, with $R_A = 2$ ohms, $L_A = 0.8$ millihenries. Calculate the fundamental current, and the current at these three harmonics.

d. Calculate the total RMS value of this current.

e. Compare the ratios of the fundamental voltage to total RMS voltage, and the ratio of fundamental current to total RMS current.

4. A round rotor permanent magnet synchronous motor has the following characteristics:

Base (rated) speed = 1350 rpm
Poles = 10
P_{rated}(continuous) = 5 kW
T_{rated}(continuous) = 35.5 Nm
$R_A = 0.01782$ ohms
$L_A = 0.301$ millihenries

Rated current = 60.5 A
Rated voltage = 54.4 V/31.4 V wye
$K_{pm} = 0.039$
(where $E_{af} = K_{pm} * \omega_e$)

$$P_{\text{mfloss}} = (0.12 \cdot I_A^2 + 0.21 * I_A + 24)\left(\frac{n}{n_{\text{base}}}\right)^2$$

The motor is operating at 500 rpm, and producing 55 N m of torque.

a. Find the rated frequency of the generator.
b. Find the torque producing current I_T.
c. Find the terminal voltage VA (remember that the flux producing current I_F will be zero since we are below base speed).
d. Find the input power to the machine.
e. Find the output power delivered by the machine.
f. Calculate the motor efficiency.

5. This same motor is now operating at 2000 rpm, above base speed. The flux producing current is $I_F = 40A$. The torque producing current is $I_T = 20A$.
   a. Find the torque produced by the motor at this operating point.
   b. Find the stator current I_A and the terminal voltage V_A. Are both of these values less than their respective rated values?
   c. Find the input and output powers for this operating point.
   d. Find the motor efficiency at this operating point.
6. This motor will not be able to operate above the point where the flux producing current equals that rated stator current. What is this speed?
7. The file induction_motor_drive_constant_torque.xls (available on request, see Preface) simulates an induction motor fed by a variable frequency inverter and operating in the constant torque region. In this file, the slip frequency is held constant while the motor speed increases from standstill to rated speed.
   The motor terminal voltage is $V_t = V_{offset} + VPH \cdot \frac{f_1}{f_{rated}}$. This is referred to as a constant volts per hertz control, with V_{offset} intended to compensate for the voltage drop across the stator resistance R_1. There are some small induction motor drives that use this control rather than field oriented control
   a. With $V_{offset} = 0$, and a slip frequency of 2 Hz, plot the stator current I_1 and the developed torque T_e versus rotor speed. Ideally, these should both be constant. Are they?
   b. Experiment with the parameter V_{offset}, in order to get a more constant value of T_e as speed varies. What value of V_{offset} gives the most constant result for T_e? How does stator current vary with speed at this value of V_{offset}?
8. A permanent magnet synchronous motor has the following ratings: 3 phase, 2 pole, 80 Hz, 480/277 V wye. $I_{rated} = 200$ A. Motor parameters are:

$$R_A = 0.20 \text{ ohms}$$
$$L_A = 0.4 \text{ millihenries}$$
$$K_{PM} = 0.46$$

   a. What is the rated speed of the motor?
   b. Using software such as MATLAB or Excel: for a range of speeds from near zero to rated speed, and with a phase current of magnitude 200 A and in phase with the internal voltage E_{af}, calculate the line to neutral motor terminal voltage V_A, the converted power P_{conv}, and the developed torque T_e.
9. The same motor is operating above rated speed at 100 Hz. The motor controller asks for a flux producing current of $I_{Aflux} = 120A$
   a. Calculate the maximum allowable torque producing current that keeps the A phase current magnitude under 200 A.
   b. For this stator current, calculate the motor terminal voltage.
   c. Is this voltage above or below rated line to neutral voltage?
   d. Calculate the converted power P_{conv}, and the developed torque T_e
10. Repeat example 8.7 with an synchronous inductance of 3 millihenries.

Lightning Source UK Ltd.
Milton Keynes UK
UKHW05n1614231018
331042UK00004BB/134/P

9 780750 316606